Communications in Asteroseismology

Volume 159
March, 2009

Proceedings of the
JENAM 2008 Symposium Nº 4: Asteroseismology and Stellar Evolution
Vienna, 8 – 12 September, 2008

edited by Sonja Schuh & Gerald Handler

Austrian Academy
of Sciences Press

Vienna 2009

OAW

Communications in Asteroseismology

Editor-in-Chief: **Michel Breger**, michel.breger@univie.ac.at
Editorial Assistant: **Daniela Klotz**, daniela.klotz@univie.ac.at
Layout & Production Manager: **Paul Beck**, paul.beck@univie.ac.at

CoAst Editorial and Production Office
Türkenschanzstraße 17, A - 1180 Wien, Austria
http://www.univie.ac.at/tops/CoAst/
Comm.Astro@univie.ac.at

Cover Illustration

Conference poster of the Joint European and National Astronomy Meeting 2008 showing the *Maria-Theresien-Statue* in front of the *Kunsthistorisches Museum* in Vienna, against a background of a galaxy field in Fornax as seen by HST's Advanced Camera for Surveys. (Poster created by Stefan Hirche, and adapted for CoAst by Katrien Kolenberg.)

British Library Cataloguing in Publication data.
A Catalogue record for this book is available from the British Library.

Austrian Academy of Sciences Press
A-1011 Wien, Postfach 471, Postgasse 7/4
Tel. +43-1-515 81/DW 3402-3406, +43-1-512 9050
Fax +43-1-515 81/DW 3400
http://verlag.oeaw.ac.at, e-mail: verlag@oeaw.ac.at

Preface

The Joint European and National Astronomy Meeting 2008 (JENAM 2008) was held from September 8-12, 2008 in Vienna, Austria, as the joint meeting of the Austrian Society of Astronomy and Astrophysics (ÖGAA), the Astronomische Gesellschaft (AG), and the European Astronomical Society (EAS). It hosted nine symposia under the overall topic "New Challenges To European Astronomy". This special volume of Communications in Asteroseismology holds the proceedings of the JENAM 2008 Symposium No 4: "Asteroseismology and Stellar Evolution". The Asteroseismology and Stellar Evolution Symposium has been generously sponsored by the ÖGAA, by the HELAS Forum (an activity of the European Helio- and Asteroseismology Network, initiative funded by the European Commission since April 1st, 2006, as a Co-ordination Action under its Sixth Framework Programme, FP6), and by the Kulturabteilung der Stadt Wien (Magistratsabteilung 7).

The Symposium program was put together by the Scientific Organizing Committee consisting of Conny Aerts (University of Leuven, Belgium), Annie Baglin (Observatoire de Paris, France), Wolfgang Glatzel (University of Göttingen, Germany), Gerald Handler (University of Vienna, Austria, Co-convener), Uli Heber (University of Erlangen-Nürnberg, Germany), Katrien Kolenberg (University of Vienna, Austria), Suzanna Randall (European Southern Observatory) and Sonja Schuh (University of Göttingen, Germany, Convener). Out of a total of 60 oral and poster contributions, 15 contributed talks were selected for presentation during the three half-day session program, with ample time for questions and discussion. There also was an opportunity to introduce the poster contributions during 2-min oral presentations.

The scientific topics covered diverse object classes divided into *Stochastically excited pulsators (main sequence, red giants, AGB supergiants)*, *Heat-driven pulsators along the main sequence*, and *Compact pulsators*, as well as additional topics that we have summarized as contributions on *Eruptive variable and binary stars*, and contributions presenting *Methods and tools*. The three session topics were introduced by two invited review speakers per field. In the first session, Jadwiga Daszynska-Daszkiewicz summarized current "Challenges for stellar pulsation and evolution theory", and Konstanze Zwintz reported "A preliminary glimpse on CoRoT results and expectations" (for Eric Michel), together giving an overview of the observational and theoretical status of the field. Anne Thoul and Oleg Kochukhov presented the state of the art in "Asteroseismology of B stars" and "Asteroseismology of chemically peculiar stars", respectively, in the second session, highlighting the hot issues in understanding and interpreting the pulsational behaviour of these groups. The stellar evolution aspect was given special consideration in "Asteroseismology and evolution of EHB stars" by Roy Østensen and "Asteroseismology and evolution of GW Vir stars" by Pierre-Oliver Quirion in the concluding third session.

A dedicated audience (on average 45 participants in each session), the excellent speakers, and our colleague Patrick Lenz who acted as the friendly and competent technician made this Symposium a highly interesting, enjoyable and successful event. We would like to thank all speakers, poster authors, and participants for coming to Vienna, the SOC and the referees for their work, the local organizers at Vienna for providing all the necessary infrastructure, and all sponsors for having made possible this event and the proceedings at hand.

Sonja Schuh and Gerald Handler
Proceedings Editors

Stochastically excited pulsators (main sequence, red giants, AGB)

Heat-driven pulsators along the main sequence

Compact pulsators

Eruptive variable and binary stars

Methods and tools

Cocktail reception at Vienna City Hall.

Comm. in Asteroseismology,
Vol. 159, 2009, JENAM 2008 Symposium № 4: Asteroseismology and Stellar Evolution
S. Schuh & G. Handler

Challenges for stellar pulsation and evolution theory

J. Daszyńska-Daszkiewicz

Instytut Astronomiczny, Uniwersytet Wrocławski, Kopernika 11, 51-622 Wrocław, Poland

Abstract

During the last few decades, great effort has been made towards understanding hydrodynamical processes which determine the structure and evolution of stars. Up to now, the most
stringent constraints have been provided by helioseismology and stellar cluster studies. However, the contribution of asteroseismology becomes more and more important, giving us an
opportunity to probe the interiors and atmospheres of very different stellar objects. A variety of pulsating variables allows us to test various parameters of micro- and macrophysics by
means of oscillation data. I will review the most outstanding key problems, both observational
and theoretical, of which our knowledge can be improved by means of asteroseismology.

Introduction

Stars are the main components of the visible Universe. Our understanding of their internal
structure and the way they evolve is a crucial piece in understanding how galaxies form
and evolve. Studies of stellar clusters bring general information about the composition and
evolution of stars but they are not sensitive enough to teach us about microphysical processes
determining stellar structure. The only particles that carry information about the solar centre
are neutrinos, which are a direct byproduct of nuclear fusion reactions. As a detection of these
weakly interacting particles is very difficult, and for stars other than the Sun still beyond our
reach, pulsations provide the only opportunity for testing the physics of stellar interiors and,
in a next step, theory of stellar evolution.

Helioseismology has by far the greatest contribution to theory of stellar structure and evolution. The global studies led to determinations of the solar age, the depth of the convective
zone, helium abundance and rotational profile. The great impact of helioseismology concerns
also atomic physics, exemplified by its role in solving the solar neutrino problem, or by testing
opacity data and equation of state. Now a new era is opening up for local helioseismology,
which should provide three-dimensional maps of the solar interior and magnetic field.

Pulsating variables cover a wide range of masses and every stage in stellar evolution. As a
consequence we can observe pulsations with various periods, amplitudes and shapes of light
curves which result from excitation of different modes. In spite of this variety, there are only
two underlying mechanisms for driving stellar pulsations. The first is self excitation in the
layers which operate as a heat engine. This instability mechanism excites pulsations in most
stars, beginning from classical instability strip stars, through B type main sequence stars, hot
subdwarfs to white dwarfs. The second way to make a star pulsate is to force stochastic
oscillations by turbulent convection. This stochastic excitation drives solar-like oscillations,
including those observed in the Sun, and is expected in all stars with extended convective
outer layers. The most complete version of the Hertzsprung-Russell diagram of pulsating

stars is shown in Fig. 1. This diagram was constructed by Simon Jeffery on the basis of an idea by Jørgen Christensen-Dalsgaard.

The diversity of stellar pulsations offers an opportunity to probe various physical phenomena, like element mixing, opacity, efficiency of convection, magnetic field and non-uniform rotation. The ultimate goal of asteroseismology is to construct a seismic model which reproduces all observed frequencies and the corresponding pulsational mode characteristics. During the last few years we are witnessing a growing impact of asteroseismology in extracting constraints on stellar parameters and physical processes in the stellar interiors. Another door to the application of asteroseismology is opening by pulsating stars which harbour planets, like μ Ara, a solar-like star (Bazot et al. 2005) or V391 Peg, a hot subdwarf (Silvotti et al. 2007).

In the first section, I will list observational key problems, which I consider most challenging for theory of stellar evolution and pulsation. The second section is devoted to the most puzzling theoretical aspects. Conclusions end this review. Because of the space limit, I had to make a crude selection. Therefore, I did not discuss, for example, the problems connected with the presence of magnetic field and the potential of asteroseismology to test it. These issues can be found, e.g., in a review by Kochukhov (2009) of the roAp pulsators.

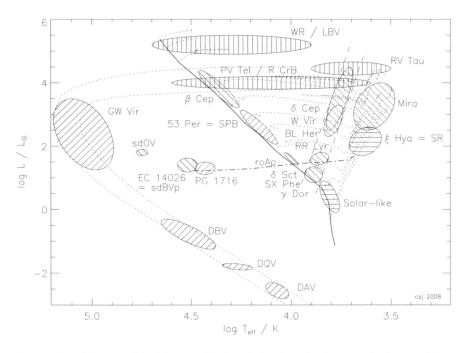

Figure 1: The Hertzsprung-Russell diagram with schematic locations of various types of pulsating variables. The hatching styles correspond to different types of stellar pulsation: diagonal lines — heat-driven p- and g-mode pulsations, horizontal lines — solar-like oscillations, and vertical lines — strange mode (highly non-adiabatic) pulsations. The zero age main sequence is depicted by a heavy solid line and the zero age horizontal branch by a heavy dot-dash line. Evolutionary tracks and white dwarf cooling sequences are shown as dotted lines. Figure courtesy of Simon Jeffery (2008a).

Observational challenges

Classical Cepheids

Classical Cepheids play a crucial role in astronomy and their importance has been recognized a long time ago. Firstly, they are primary distance indicators on the extragalactic scale thanks to the Period Luminosity Relation discovered by Henrietta Leavitt in 1908. Secondly, their pulsations provide a test of stellar evolution theory of intermediate mass stars.

The longest standing problem connected with these variables was a discrepancy for double-mode Cepheids between masses estimated from pulsation and evolution theory. The implementation of new opacity data (Iglesias & Rogers 1991) made by Moskalik et al. (1992) allowed essentially to reconcile this mass discrepancy. Despite this crucial work some disagreement still remains and possible sources of this persistent problem are opacities, mass loss and internal mixing processes (Keller 2008 and reference therein). Recently, yet another problem arose for double-mode Cepheid models. Smolec & Moskalik (2008a, b) have shown that if buoyancy forces in convectively stable layers are included in pulsation modelling, double-mode solutions do not exist.

During the last few years a number of interesting and challenging discoveries were made, based mostly on the OGLE project data. There are:
- Blazhko Cepheids (Moskalik & Kołaczkowski 2008a)
- non-radial modes in classical Cepheids (Moskalik & Kołaczkowski 2008b)
- 1O/3O double-mode Cepheids (Soszyński et al. 2008a)
- single-mode second-overtone Cepheids (Udalski et al. 1999, Soszyński et al. 2008b),
- triple-mode Cepheids (Moskalik et al. 2004, Soszyński et al. 2008a, b)
- eclipsing binary systems containing Cepheids (Soszyński et al. 2008b).

All the above theoretical and observational facts wait for explanation and call for new generation models of stellar pulsation and evolution.

B type main sequence stars

Studying B type main sequence stars is of great importance because those with masses greater than about $8\,M_\odot$ are progenitors of Type II Supernovae, whereas those with masses less than about $8\,M_\odot$ form the CNO elements in the Universe. In some of these objects pulsation occurs, giving an opportunity to study physics of stellar interiors by means of oscillation data. There are two classes of B type main sequence pulsators: 1) β Cephei stars with masses larger than $8\,M_\odot$ and spectral types B0–B2.5, in which mainly pressure (p) modes are excited, and 2) Slowly Pulsating B type (SPB) stars with masses smaller than $8\,M_\odot$ and spectral types B3–B9, which pulsate in high-order gravity (g) modes.

Many years had elapsed since the discovery of β Cephei stars before the cause of their pulsations was identified. As in the case of the classical Cepheid mass discrepancy problem, the reason was that the Los Alamos opacities missed the metal bump because of underestimations of the heavy element contributions, mostly from the iron group (e.g., Moskalik & Dziembowski 1992, Dziembowski et al. 1993). As pulsations in β Cephei and SPB stars are strictly connected with the metal opacity bump, nobody expected to find many of them in the Magellanic Clouds (MCs), because the metallicity in the MCs is much lower than in the Milky Way; $Z = 0.007$ for the Large Magellanic Cloud and $Z = 0.002$ for the Small Magellanic Cloud. Therefore, the discovery of a large number of β Cep and SPB variables in the MCs was rather a surprise. This started with the pioneering paper by Pigulski and Kołaczkowski (2002). The later papers by Kołaczkowski et al. (2006) and Karoff et al. (2008) increased the number of known B type main sequence pulsators in the MCs. Moreover, the newly determined solar chemical composition (Asplund et al. 2005, AGS05) gives a lower metallicity of about $Z = 0.012$ compared to the older one (Grevesse & Noels 1993) $Z = 0.017$. Nevertheless, the new Z value does not cause disappearance of the B main sequence instability strip, as one would initially expect, because most of the reduction was obtained in the CNO elements and now the relative abundances of Fe group elements are significantly higher than the older ones (Pamyatnykh & Ziomek 2007, Miglio et al. 2007). However, the problem with the existence of β Cep stars in SMC remains open. This observational fact waits for an explanation and indicates that improvements are still needed in the treatment of opacities, mixing processes, diffusion etc., as well as more observational data should be gathered.

Another challenge is the occurrence of hybrid B type pulsators like ν Eri and 12 Lac (e.g., see Handler 2009). The problem is that the lower frequencies observed in the g mode range cannot reach the instability in pulsational models. This instability problem occurs also for very high-frequency modes. These two stars were recently studied by Dziembowski & Pamyatnykh (2008), who highlighted problems of mode excitation, uncertainties in opacity data and element distribution, extent of the overshooting distance and internal rotation.

Hot subdwarfs

B type subdwarfs are stars in the core helium burning phase with a thin hydrogen envelope. No hydrogen shell burning takes place. These stars represent the final stage before the white dwarf phase but only 2% of white dwarfs are formed through this channel.

Short period pulsations in these objects were theoretically predicted by Charpinet et al. (1997); Kilkenny et al. (1997) found them observationally. Then, Green et al. (2003) found B type subdwarfs pulsating with long periods. Both types of pulsation are driven by the κ mechanism due to the Z-opacity bump. Charpinet et al. (1997) incorporated a nonuniform iron profile as determined by the condition of diffusive equilibrium between gravitational settling and radiative levitation. This allowed removing the artificial assumption of a very high value of the metallicity. The present status of the hot subdwarfs asteroseismology was summarized by Fontaine et al. (2008a). The evolutionary status of subluminous B type stars seems to be proven but not their origin. There are two most supported scenarios: 1) single star evolution, or 2) binary star evolution, which can be a common envelope evolution, a stable Roche lobe overflow or a merger of two He WD stars.

Recently, also pulsation in O type subdwarfs was discovered (Woudt et al. 2006, Rodriguez-Lopez et al. 2007). The sdO stars are the more evolved and even more puzzling cousins of sdB stars. These objects have a C/O core and are in the helium burning shell phase. As they cover quite a large part of the HR diagram, they are thought to have many different origins. The "luminous" (low gravity) sdO stars are believed to be post-AGB stars. For the "compact" (high gravity) sdO stars two origin scenarios were proposed depending on the helium abundance: 1) post EHB objects (descendants of sdBs) for He-deficient sdOs, and 2) a merger of two He WDs or a delayed core He flash scenario for He-enriched sdOs (Rodriguez-Lopez et al. 2007). Fontaine et al. (2008a) performed non-adiabatic pulsational calculations taking into account time-dependent diffusion, as has been done before for sdB pulsators. They showed that radiative levitation causes pulsational instabilities in the sdO stars through an iron accumulation in the driving region.

There is great potential for asteroseismology of OB subdwarf pulsators because their radiative atmospheres combined with high gravities make them ideal for investigating diffusive processes. The problem with each origin scenario of sdOB stars is that a large amount of hydrogen has to be lost before or just at the beginning of helium core ignition. As in the case of the main sequence B type pulsators, additional opportunities come from hybrid sdOB pulsators (Schuh et al. 2006, Lutz et al. 2008).

Extreme helium stars

The extreme helium stars are low-mass highly evolved objects (supergiants) of B and A types, with spectra showing very weak or no hydrogen lines (Jeffery 2008b). They cover a wide range of effective temperature and most of them have luminosities close to the Eddington limit. The fundamental questions about this kind of stars are their origin (how to remove a hydrogen-rich envelope) and their connection with stars with normal He abundance. Two main origin scenarios were proposed. In the first scenario, the EHe star is a direct product of a single white dwarf which underwent a late thermal pulse (e.g., Iben et al. 1983). The second scenario involves a white dwarf binary which has merged (e.g., Webbink 1984).

The existence of EHe stars exhibiting pulsational variability is extremely significant. Jeffery (2008c) introduced consistent classification of variable EHe stars. The first group consists of PV Tel stars with strange mode pulsational instability (Saio & Jeffery 1988) which is supposed to be present in all stars with sufficiently high luminosity/mass ratio. The second class includes BX Cir stars with pulsations driven by the κ mechanism operating in the Z bump (Saio 1993). Here, the role of the iron group element opacity is increased due to a reduction of hydrogen (Jeffery & Saio 1999).

There are also helium-rich subluminous OB stars whose origin is unclear (Napiwotzki 2008, Ahmad & Jeffery 2008), but the hypothesis of merging two He core white dwarfs seems to be preferred. Questions arise about their connection with hydrogen-rich sdOB stars or with EHe stars. In the light curve of one He sdB object, Ahmad & Jeffery (2005) detected multiperiodic variations. These authors have shown that the variations can be associated with high-order g-mode pulsations, but according to theory such modes are stable.

New class of white dwarfs

Asteroseismology of white dwarf (WD) pulsators provides many constraints on stellar physics and evolution. For example, from period changes we can estimate the cooling rate, which in turn can give information on the age of our Galaxy and can measure the neutrino generation rate in hot WDs (Winget 1998). WD asteroseismology also supplies a unique test of the equation of state of matter at high densities and temperatures. Because most stars (97%) will end up as WDs, their studies are of special importance. An excellent review on the three types of WD pulsators, i.e., ZZ Cet (DAV), V777 Her (DBV) and GW Vir (PNNV+DOV) type stars, can be found in Fontaine & Brassard (2008).

WDs of DA spectral type have hydrogen-rich atmospheres. The atmospheres of the two other types (DB, DO+PNNV) are dominated by helium. There is one more class of WDs at T_{eff} 11000 − 13000 K, with carbon lines in their spectra, designated the DQ type, but until recently even in this class, helium was believed to be the dominant component of the atmosphere. Therefore, the discovery of hot DQ WDs (T_{eff} 18000 − 23000 K) with carbon-dominated atmospheres and little or no H and He, was rather unexpected (Dufour et al. 2007). The origin of the cooler DQ WDs is quite well understood in the framework of the model of carbon dredge-up by the deep helium convection zone. On the other hand, the origin of hot DQ WDs cannot be explained within any known post-AGB evolution channel.

Soon thereafter, Fontaine et al. (2008b) have explored the instability of hot DQ WD models against pulsation. Their analysis showed that gravity modes should be excited in such models in the period range 100 − 700 s. Almost at the same time, Montgomery et al. (2008) announced the photometric variability of one carbon-atmosphere object with a period of 417 s. This makes this object a prototype of a new class of pulsating WDs. Recently, two more variable DQ WDs have been discovered by Barlow et al. (2008). Having in mind the potential of asteroseismology, one can expect interesting results from seismic studies of hot DQ WD pulsators before long.

Theoretical challenges

Opacities and chemical composition

Stellar opacities constitute a vital component in modelling stellar structure because they determine the transport of radiation through matter. Their values depend on the temperature, density and chemical composition, $\kappa(T, \rho, X_i)$. For a long time the only source of the opacity data for astrophysical purposes was the Los Alamos Opacity Library (LAOL). The LAOL was in use until 1990, although already in 1982 N. R. Simon dared to blame them for the failure in explaining the Cepheid mass discrepancy and pulsations of β Cephei stars, and urged reexamination of this crucial input (Simon 1982). Finally, in early 90ties, two teams of atomic physicists recomputed the opacity data and found that the LAOL underestimated the contributions from the heavy elements by a factor of $2-3$ at a temperature of about 200 000 K. The first team was represented by Iglesias & Rogers (e.g., 1991, 1996), who called their results OPAL (OPAcity Library). The second was an international team led by M. J. Seaton and their opacities were named OP (Opacity Project, e.g., Seaton 1992, 1996, 2007).

The computation of new opacity tables was a milestone for astrophysics and the main consequences of a huge enhancement of the metal opacity bump were: 1) the seismic model of the Sun was improved, 2) the Cepheid mass discrepancy was significantly reduced, 3) the pulsations of B type main sequence stars and of 4) some extreme He stars were explained, 5) pulsations of sdB and sdO stars were predicted.

The opacity data are being constantly updated but the main features are kept unchanged. Recently, a substantial revision of the solar chemical mixture has led to a significant change of the opacity values (AGS05). The new solar composition was reduced mostly in the CNO elements and now the solar metallicity is only 70% of the older one by Grevesse & Noels (1993). As mentioned before, these new solar abundances did not diminish the extent of the pulsational instability strip of B type stars because the relative Fe abundance was increased, hence the role of the Z bump opacity in driving pulsations was amplified. There is other strong evidence supporting the AGS05 mixture. For example, now the solar metallicity is in better agreement with metallicities of stars in its neighbourhood. Then, galactic beat Cepheid models with the AGS05 abundances better fit the observations (Buchler & Szabo 2007). Moreover, with reduced Z, the pre-main sequence models show smaller lithium depletion (Montalbán & D'Antona 2006) which brings them closer to the observations. However, the new solar composition had bad consequences for the helioseismic model which lost consistency with the standard solar model. This problem was extensively discussed in a review by Basu & Antia (2008).

There are many papers in which stellar pulsations of various objects were used as a test of opacity data, e.g., Dziembowski & Pamyatnykh (2008) for β Cep stars, Jeffery & Saio (2006) for pulsating subdwarf B stars, Lenz et al. (2008) for δ Sct stars or Théado et al. (2009) for roAp stars. The role of input from opacity data in pulsation computations was recently summed up by Montalbán & Miglio (2008).

Rotation

A credible theory of stellar structure and evolution should incorporate rotation. Firstly, rotation affects the stellar structure by breaking spherical symmetry. The most extreme example known is Achernar with a ratio of the major to minor axes equal to 1.56 ± 0.05. Secondly, rotation activates various processes, like meridional circulation, shear instabilities, diffusion, horizontal turbulence, which cause mixing. Thirdly, the distribution of internal angular momentum is determined through different rotational velocity at different depths. And finally, mass loss from the surface can be enhanced by rapid rotation through centrifugal effects. Many excellent papers on stellar rotation have been published by, e.g., Sweet, Öpik, Tassoul, Roxburgh, Zahn, Spruit, Deupree, Talon, Meynet, Maeder, Mathis, but there is no space here to mention the results of all of them. In the last years the most comprehensive studies of the rotational effects on stellar structure and evolution were presented in a series of papers by Maeder, Meynet and their collaborators (e.g., Maeder & Meynet 2000, Meynet & Maeder 2000).

How rotation affects pulsation depends on the rotation rate and on the closeness of oscillation frequencies. If the pulsational frequency is much larger than the rotational angular velocity, the perturbation approach is applicable; this considerably simplifies calculations and in the first-order approximation each pulsational mode can be described by a single spherical harmonic. This is no longer true if the higher-order effects of rotation are included (Dziembowski & Goode 1992). Moderate rotation can also couple modes for which the frequency distance is close to the rotational angular velocity and where the spherical harmonic indices satisfy the relations: $l_j = l_k + 2$ and $m_j = m_k$ (Soufi et al. 1998). The third-order expression for a rotationally split frequency can by found in Goupil et al. (2000). As for mode geometry, the main effect of fast rotation is the confinement of pulsation towards the stellar equator (e.g., Townsend 1997).

An entirely different approach has to be applied if the pulsation frequencies are of the order of or smaller than the rotation frequency ($\nu_{puls} \sim \nu_{rot}$). Three different treatments of slow modes can be found in the literature: the traditional approximation (Townsend 2003), expansion in Legendre function series (Lee & Saio 1997) and 2D(r, θ) modelling (Savonije 2007).

Rotation also complicates the identification of pulsation modes from photometric diagnostic diagrams because they become dependent on inclination angle, i, azimuthal order, m, and rotation velocity, v_{rot}. This problem was studied by Daszyńska-Daszkiewicz et al. (2002) for coupled modes, and by Townsend (2003) and Daszyńska-Daszkiewicz et al. (2007) for slow modes.

The asteroseismic potential of rotating pulsators lies in the rotational splitting kernel, $K(r)$ that gives information on the rotational profile $\Omega(r)$. The result that the rotation rate increases inward was obtained from studies of many pulsating stars, e.g., Goupil et al. (1993) for the δ Sct star GX Peg; Dziembowski & Jerzykiewicz (1996) for the β Cep star 16 Lac; Aerts et al. (2003) for the β Cep star V836 Cen, Pamyatnykh et al. (2004) for the β Cep star ν Eri and Dziembowski & Pamyatnykh (2008) for the β Cep stars ν Eri and 12 Lac.

Another question is about the impact of pulsation on rotational evolution. Talon & Charbonnel (2005) showed that internal gravity waves contribute to braking rotation in the inner regions of low-mass stars. Computations by Townsend & MacDonald (2008) demonstrated that pulsation modes can redistribute angular momentum and trigger shear instability mixing in the μ gradient zone. Mathis et al. (2008) discussed the transport of angular momentum in the solar radiative zone by internal gravity waves.

Convection

Convection plays a very important role in the transport of energy and mixing of matter. It is a complex physical process with a three-dimensional, non-local and time-dependent character. The convective cells are also a source of acoustic waves in subphotospheric layers, leading to stochastic excitation of stellar oscillations. In turn, dissipation of acoustic energy heats stellar chromospheres, causing "fingerprints" in spectral lines. Also stellar activity is a result of joint action of convection and differential rotation.

The most widely used description of stellar convection is the Mixing Length Theory (MLT) due to Böhm-Vitense (1958) and some modifications of it. In the framework of MLT, the size of convective elements is parameterized by the the mixing length parameter, α_{conv}, which is adjusted to fit some observational quantities. A step forward was done by Canuto et al. (1996), who formulated a theory of turbulent convection taking into account the full spectrum of convective eddies. A 3D hydrodynamical simulation by Stein & Nordlund (1998) allowed to reproduce qualitatively convection in the solar surface layers.

A discussion of differences between the 3D hydrodynamical model of convection and 1D MLT model was presented by Steffen (2007). The main results of his comparisons were that it is impossible to reproduce the correct temperature profile with any value of the MLT parameter, and that the radiative layer between two convection zones is completely mixed.

The presence of pulsation in stars with expanded convective envelopes complicates the picture even more. This is the case for classical Cepheids, RR Lyrae stars, red giant pulsators, δ Scuti and γ Doradus stars, as well as for pulsating white dwarfs of the V777 Her and ZZ Cet types. A simplistic approach to pulsational modelling of these objects is a convective flux freezing approximation which assumes that the convective flux is constant during the pulsation cycle. Although in some δ Scuti stars convection seems to be inefficient (Daszyńska-Daszkiewicz et al. 2003, 2005a), this is a crude approximation and an adequate treatment of pulsation-convection interaction should be applied.

The first formulation of pulsation-convection interactions in which convection is non-local and time-dependent was given by Unno (1967) and Gough (1977). Following this concept and its variants, many research groups undertook attempts to model stellar oscillations in various objects: solar-like stars (Houdek, Goupil, Samadi), δ Scuti and γ Doradus stars (Xiong, Houdek, Dupret, Grigahcene, Moya), classical Cepheids and RR Lyr variables (Feuchtinger, Stellingwerf, Buchler, Kollath, Smolec, Moskalik), pulsating red giants (Xiong, Deng, Cheng) and recently, V777 Her (DBV) white dwarfs (Quirion, Dupret).

The potential of asteroseismology for estimating the size of the convective core in massive stars was also explored. Dziembowski & Pamyatnykh (1991) demonstrated that modes which are largely trapped in the region surrounding the convective core boundary can measure the extent of overshooting. The first evidence of core overshooting in a β Cep star was found by Aerts et al. (2003) for V836 Cen. Miglio et al. (2008) studied the sensitivity of high-order g modes in SPB and γ Dor stars to the properties of convective cores. In particular, the period spacing of gravity modes depends on the location and shape of the chemical composition gradient.

In general, convective transport of energy above the core in OB type stars is negligible. However, recently Maeder et al. (2008) have shown that very fast rotation in massive OB stars can increase the extent of outer convective envelopes. This has many consequences, e.g., acoustic modes can be generated. Also, stars close to the Eddington limit may develop a convective envelope.

Mass loss

Mass loss occurs in all late evolutionary phases and in massive stars. There are two principal mechanisms driving stellar winds: radiation driving in hot stars and dust driving in cool and luminous stars (e.g., Owocki 2004). For most stars, no consistent description of the mass loss rate exists and empirical formulae are used instead. Also, the interaction between mass loss and rotation has not yet been fully understood (e.g., Maeder & Meynet 2004, Owocki 2008). Some stars with strong winds exhibit also pulsations. These are Mira and semi-regular (SR) variables, massive OB main sequence stars, Wolf-Rayet stars and Luminous Blue Variables (LBV). For such objects a natural question emerges about pulsation and mass loss coupling. It has been recognized many years ago that pulsations in large amplitude variables, like Miras or Cepheids, can enhance mass loss (e.g., Wood 2007, Neilson & Lester 2008). Constraints have been derived from relations between the mass loss rate and pulsational period and between the wind outflow speed and pulsational period (e.g., Knapp et al. 1998). As for hot pulsators, Howarth et al. (1993) found wind variability in ζ Oph with the pulsational cycle, and Kaufer et al. (2006) detected the pulsation beat period in Hα profile observations for a B0 type supergiant. A nice discussion of the coupling between mass loss and pulsation in massive stars can be found in Townsend (2007).

Interesting results have been obtained recently by Quirion, Fontaine & Brassard (2007) for hot white dwarf pulsators of the GW Vir type. The authors showed that useful constraints on mass loss can be inferred from the red edge position of the instability strip.

Conclusions

An ultimate goal of asteroseismology is to help solving the equation *observation = theory* and to avoid the equation *more data = less understanding*. But as we are all aware, a more realistic treatment of macro- and microphysics in stellar modelling would be desirable.

One should have in mind that not only pulsation frequencies can probe stellar structure. An ideal seismic stellar model should account both for all measured oscillation frequencies and for associated pulsation mode characteristics. Therefore more use should be made of photometric and spectroscopic observables, i.e., amplitudes and phases of photometric and spectroscopic variations, and simultaneous multi-colour photometric and spectroscopic observations should be carried out. From such data we can infer an additional asteroseismic quantity, which is the ratio of the bolometric flux variations to the radial displacement. This new asteroseismic probe, called the f parameter, is determined in subphotospheric layers, and therefore it is complementary to the frequency data which poorly probe these stellar regions. The value of f is very sensitive to global stellar parameters, element mixture (hence the mixing processes), opacities and subphotospheric convection. Therefore, a comparison of empirical and theoretical values of f provides additional stringent constraints on various physical parameters and processes. Such asteroseismic studies were proposed by Daszynska-Daszkiewicz et al. (2003, 2005ab) and successfully applied to δ Scuti and β Cephei stars. In the case of δ Scuti stars, useful constraints on subphotospheric convection were derived, and in the case of β Cep stars, on opacities.

In the next step, the f parameter should be fitted to observations together with the pulsation frequency. These complex asteroseismic studies should lead to stronger constraints, improving our knowledge in theory of stellar pulsation and evolution.

Acknowledgments. The author would like to thank the SOC members of the JENAM 2008 Symposium No. 4 for their invitation and Mikołaj Jerzykiewicz and Alosza Pamyatnykh for carefully reading the manuscript and their comments. The EC is acknowledged for the establishment of the European Helio- and Asteroseismology Network (HELAS, No. 026138), which made the authors' participation at this meeting possible.

References

Aerts, C., Thoul, A., Daszyńska, J., et al. 2003, Science, 300, 1926

Ahmad, A., & Jeffery, C. S. 2005, A&A, 437, L51

Ahmad, A., & Jeffery, C. S. 2008, in "Hydrogen-Deficient Stars", eds. K. Werner and T. Rauch, ASP Conf. Ser., 391, 261

Asplund, M., Grevesse, N., Sauval, A. J., et al. 2005, A&A, 431, 693

Barlow, B. N., Dunlap, B. H., Rosen, R., & Clemens, J. C. 2008, ApJ, 2008, 688, L95

Basu, S., & Antia, H. M. 2008, Phys. Rep., 457, 217

Bazot, M., Vauclair, S., Bouchy, F., & Santos, N. C. 2005, A&A, 440, 615

Böhm-Vitense, E. 1958, Z. Astrophys., 46, 108

Buchler, J. R., & Szabo, R. 2007, ApJ, 660, 723

Canuto, V. M., Goldman, I., & Mazzitelli, I. 1996, ApJ, 473, 550

Charpinet, S., Fontaine, G., Brassard, P., et al. 1997, ApJ, 483, L123

Daszyńska-Daszkiewicz, J., Dziembowski, W. A., Pamyatnykh, A. A., & Goupil, M.-J. 2002, A&A, 392, 151

Daszyńska-Daszkiewicz, J., Dziembowski, W. A., & Pamyatnykh, A. A. 2003, A&A, 407, 999

Daszyńska-Daszkiewicz, J., Dziembowski, W. A., Pamyatnykh, A. A., et al. 2005a, A&A, 438, 653

Daszyńska-Daszkiewicz, J., Dziembowski, W. A., & Pamyatnykh, A. A. 2005b, A&A, 441, 641

Daszyńska-Daszkiewicz, J., Dziembowski, W. A., & Pamyatnykh, A. A. 2007, AcA, 57, 11

Dufour, P., Liebert, J., Fontaine, G., & Behara, N. 2007, Nature 450, 522

Dziembowski, W. A., & Goode, P. R. 1992, ApJ, 394, 670

Dziembowski, W. A., & Jerzykiewicz, M. 1996, A&A, 306, 436

Dziembowski, W. A., & Pamyatnykh, A. A. 1991, A&A 248, L11

Dziembowski, W. A., & Pamyatnykh, A. A. 1993, MNRAS, 226, 204

Dziembowski, W. A., & Pamyatnykh, A. A. 2008, MNRAS, 385, 2061

Fontaine, G., Brassard, P., Charpinet, S., et al. 2008, in "Hot Subdwarf Stars and Related Objects", eds.
    U. Heber, R. Napiwotzki, and C. S. Jeffery, ASP Conf. Ser., 392, 231

Fontaine, G., & Brassard, P. 2008, PASP, 120, 104

Fontaine, G., Brassard, P., Green, B., et al. 2008a, A&A, 486, L39

Fontaine, G., Brassard, P., & Dufour, P. 2008b, A&A, 483, L1

Green, E. M., Fontaine, G., Reed, M. D., et al. 2003, ApJ 583, L31

Gough D., 1977, ApJ, 214, 196

Goupil, M. J., Michel, E., Lebreton, Y., Baglin, A., 1993, A&A, 268, 546

Goupil, M.-J., Dziembowski, W. A., Pamyatnykh, A. A., & Talon, S. 2000, in "Delta Scuti and Related
    Stars", eds. M. Breger and M. H. Montgomery, ASP Conf. Ser., 210, 267

Grevesse, N., Noels, A. 1993, in "Origin and Evolution of the Elements", eds. N. Pratzo,
    E. Vangioni-Flam, and M. Casse, Cambridge Univ. Press, Cambridge, p. 15

Handler, G. 2009, CoAst, 159, 42

Howarth I. D., Bolton, C. T., Crowe, R. A., et al. 1993, ApJ, 417, 338

Iben, J., Kaler, J. B., Truran, J. W., & Renzini, A. 1983, ApJ, 264, 605

Iglesias, C. A., & Rogers, F. J. 1991, ApJ, 371, 408

Iglesias, C. A., & Rogers, F. J. 1996, ApJ, 464, 943

Jeffery, C. S., Drilling, J. S., Harrison, P. M., et al. 1997, A&AS, 125, 501

Jeffery, C. S., & Saio, H. 1999, MNRAS,308, 221

Jeffery, C. S., & Saio, H. 2006, MNRAS,372, L48

Jeffery, C. S. 2008a, CoAst, 157, 240

Jeffery, C. S. 2008b, in "Hydrogen-Deficient Stars", eds. K. Werner and T. Rauch, ASP Conf. Ser., 391,
    53

Jeffery, C. S. 2008c, IBVS, 5817, 1

Karoff, C., Arentoft, T., Glowienka, L., et al. 2008, MNRAS, 386, 1085

Kaufer, A., Stahl, O., Prinja, R. K., & Witherick, D., 2006, A&A, 447, 325

Keller, S. C. 2008, ApJ, 677, 483

Kochukhov, O. 2009, CoAst, 159, 61

Kołaczkowski, Z., Pigulski, A., Soszyński, I., et al. 2006, MmSAI, 77, 336

Kilkenny, D., Koen, C., O'Donoghue, D., & Stobie, R. S. 1997, MNRAS, 285, 640

Knapp, G. R., Young, K., Lee, E., & Jorissen, A. 1998, ApJS, 117, 209

Lenz, P., Pamyatnykh, A. A., Breger, M., & Antoci, V. 2008, A&A, 478, 855

Lutz, R., Schuh, S., Silvotti, R., et al. 2008, in "Hot Subdwarf Stars and Related Objects", eds.
    U. Heber, C. S. Jeffery, and R. Napiwotzki, ASP Conf. Ser., 392, 339

Maeder, A., Georgy, C., & Meynet, G. 2008, A&A, 479, L37

Mathis, S., Talon, S., Pantillon, F.-P., & Zahn, J.-P. 2008, SoPh, 251, 101

Meynet, G., & Maeder, A. 2000, A&A, 361, 101

Maeder, A., & Meynet, G. 2000, ARA&A, 38, 143

Miglio, A., Montalbán, J., & Dupret, M.-A. 2007, MNRAS, 375, L21

Miglio, A., Montalbán, J., Noels, A., & Eggenberger P. 2008, MNRAS, 386, 1487

Montalbán, J., D'Antona, F. 2006, MNRAS, 370, 1823

Montalbán J., & Miglio, A. 2008, CoAst, 157, 160

Montgomery, M. H., Williams, A. K., Winget, D. E., et al. 2008, ApJ, 678, L51

Moskalik, P., Buchler, J. R., & Marom, A. 1992, ApJ, 385, 685

Moskalik, P., & Dziembowski, W. A. 1992, A&A, 256, L5

Moskalik, P., & Kołaczkowski, Z. 2008a, arXiv:0807.0615

Moskalik, P., & Kołaczkowski, Z. 2008b, arXiv:0807.0623

Moskalik, P., Kołaczkowski, Z., & Mizerski, T. 2004, ASP Conf. Ser., 310, 498

Napiwotzki, R. 2008, in "Hydrogen-Deficient Stars", eds. K. Werner and T. Rauch, ASP Conf. Ser., 391, 257

Neilson, H. R., & Lester, J. B. 2008, ApJ, 684, 569

Owocki, S. 2004, in "Evolution of Massive Stars, Mass Loss and Winds", eds. M. Heydari-Malayeri, Ph. Stee, and J.-P. Zahn, EAS Publ. Ser., 13, 163

Owocki, S. 2008, in "Mass Loss from Stars and the Evolution of Stellar Clusters", eds. A. de Koter, L. J. Smith, and L. B. F. M. Waters, ASP Conf. Ser., 388, 57

Pamyatnykh, A. A., & Ziomek, W. 2007, CoAst, 150, 207

Pamyatnykh, A. A., Handler, G., & Dziembowski, W. A. 2004, MNRAS, 350, 1022

Pigulski, A., & Kołaczkowski, Z. 2002, A&A, 388, 88

Quirion, P. O., Fontaine, G., & Brassard, P. 2007, CoAst, 150, 247

Rodriguez-Lopez, C., Ulla, A., & Garrido, R. 2007, MNRAS, 379, 1123

Saio, H. 1993, MNRAS, 260, 465

Saio, H., & Jeffery, C. S. 1988, ApJ, 328, 714

Lee, U., & Saio, H. 1997, ApJ., 491, 839

Savonije, G. J. 2007, A&A, 469, 1057

Seaton, M. J. 1992, RMxAA, 23, 180

Seaton, M. J. 1996, MNRAS, 279, 95

Seaton, M. J. 2007, MNRAS, 382, 245

Schuh, S., Huber, J., Dreizler, S., et al. 2006, A&A, 445, L31

Silvotti, R., Schuh, S., Janulis, R., et al. 2007, Nature 449, 189

Simon, R. N. 1982, ApJ, 260, L87

Smolec, R., & Moskalik, P. 2008a, AcA, 58, 233

Smolec, R., & Moskalik, P. 2008b, AcA, 58, 193

Soufi, F., Goupil, M-J., & Dziembowski, W. A. 1998, A&A, 334, 911

Soszyński, I., Poleski, R., Udalski, A., et al. 2008a, AcA, 58, 153

Soszyński, I., Poleski, R., Udalski, A., et al. 2008b, AcA, 58, 163

Stein, R. F., & Nordlund, A. 1998, ApJ, 499, 914

Steffen, M. 2007, in "Convection in Astrophysics", eds. F. Kupka, I. Roxburgh, and K. Chan, IAU Symposium, 239, 36

Talon, S., & Charbonnel, C. 2005, A&A, 440, 981

Théado, S., Dupret, M.-A., Noels, A., & Ferguson, J. W. 2009, A&A 493, 159

Townsend, R. 1997, MNRAS, 284, 839

Townsend, R. 2003, MNRAS, 343, 125

Townsend, R. 2007, in "Unsolved Problems in Stellar Physics", AIP Conf. Proceedings, 948, 345

Townsend, R., & MacDonald, J. 2008, in "Massive Stars as Cosmic Engines", eds. F. Bresolin,
    P. A. Crowther, and J. Puls, IAU Symposium, 250, 161

Udalski, A., Soszyński, I., Szymański, M., et al. 1999, AcA, 49, 45

Unno, W., 1967, PASJ, 19, 140

Webbink, R. F. 1984, ApJ, 277, 355

Winget, D. E. 1998, J. Phys., Condensed Matter, 10, 11247

Wood, P. R. 2007, in "From Stars to Galaxies: Building the Pieces to Build Up the Universe", eds.
    A. Vallenari, R. Tantalo, L. Portinari, and A. Moretti, ASP Conf. Ser., 374, 47

Woudt, P. A., Kilkenny, D., Zietsman, E., et al. 2006, MNRAS, 371, 1397

Comm. in Asteroseismology,
Vol. 159, 2009, JENAM 2008 Symposium № 4: Asteroseismology and Stellar Evolution
S. Schuh & G. Handler

μ Herculis: Analysis of EW time series of a solar-type pulsator

E. Carolo,[1] R. Claudi,[1] S. Benatti,[2] and A. Bonanno[3]

[1] INAF - Osservatorio Astronomico di Padova, Vicolo Osservatorio 5, 35122, Padova, Italy
[2] CISAS, Universitá degli Studi di Padova, via Venezia 15, 35131, Padova, Italy
[3] INAF - Osservatorio Astrofisico di Catania, via S. Sofia 78, 95123, Catania, Italy

Abstract

The G5 subgiant star μ Herculis was observed in June 2006 by means of the high-resolution spectrograph SARG operating at the 3.6 m Italian telescope TNG (Telescopio Nazionale Galileo) at the Canary Islands, exploiting the iodine cell technique. A time series of about 1200 spectra was acquired during 7 observing nights. Data analysis of the radial velocity time series (Bonanno et al. 2008) has shown a significant power excess centred at 1.2 mHz, with ~ 0.9 ms^{-1} peak amplitude. Here we present an analysis of the Equivalent Width time series of H$_\alpha$ and other lines.

Individual Objects: μ Her

Method

We considered the red part of echelle spectrum of μ Her (between 600 and 900 nm), since it is insensitive to the iodine cell. This allows us to use the Equivalent Width (EW) variations to evaluate the presence of stellar pulsations and granulation. The analyzed lines were: Fe I (λ=6393.2, 6545.8, 6677.6 Å), Ni I (6643.25 Å) and H$_\alpha$ (6563 Å). The H$_\alpha$ line is on the rim of two adjacent spectral orders of the echelle spectrum. In order to have the red and blue wings of the line both on the same spectrum, we merged the two parts. To do this in a safe way, we correct the two orders for the continuum before merging them. For all the lines we evaluate the EW in the usual way, using the following expression:

$$EW = \int_{\lambda_1}^{\lambda_2}(1 - R_\lambda)d\lambda, \text{ where } R_\lambda = \frac{f(\lambda)}{f(c)}$$

Time series and power spectra of EWs

Figure 1 shows the EW time series of one of the analyzed lines (H$_\alpha$). The observations were conducted on 7 nights in June 2006 using SARG, the high resolution spectrograph of the TNG. As expected the computed power spectrum of the EW time series shows a very low amplitude. Figure 2 shows as example the H$_\alpha$ Power Spectrum.

The power spectrum does not show evidence of a power excess due to pulsation. Nevertheless a concentration of power is present at low frequencies. This feature is similar to the background power found by Kjeldsen et al. (1999) for α Cen A and by Leccia et al. (2007) for Procyon A. These authors hypothesized this noise to be due to stellar granulation.

A more detailed analysis of the time series of EWs and the estimate of the stellar granulation using by power density spectrum is in progress.

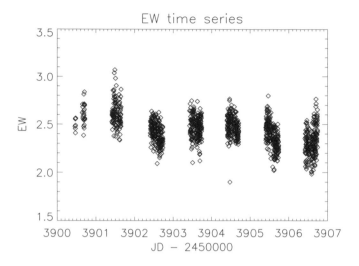

Figure 1: Time series of EWs of H_α.

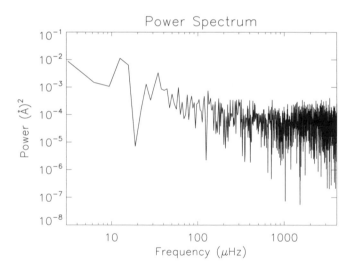

Figure 2: Power spectra of EWs of H_α.

References

Bonanno, A., Benatti, S., Claudi, R., et al. 2008, ApJ, 676, 1248

Kjeldsen, H., Bedding, T. R., Frandsen, S., & Dall, T. H. 1999, MNRAS, 303, 579

Leccia, S., Kjeldsen, H., Bonanno, A., et al. 2007, A&A, 464, 1059

Comm. in Asteroseismology,
Vol. 159, 2009, JENAM 2008 Symposium № 4: Asteroseismology and Stellar Evolution
S. Schuh & G. Handler

Asteroseismology of solar–type stars with SARG@TNG

R. Claudi,[1] S. Benatti,[2] A. Bonanno,[3] M. Bonavita,[1] S. Desidera,[1] R. Gratton,[1] S. Leccia,[4]
R. Cosentino,[5] M. Endl,[6] and E. Carolo[1]

[1] INAF - Osservatorio Astronomico di Padova, vicolo Osservatorio 5, 35122 Padova, Italy
[2] CISAS, Universitá degli studi di Padova, via Venezia 15, 35131 Padova, Italy
[3] INAF - Osservatorio Astrofisico di Catania, via S. Sofia, 78, 95123 Catania, Italy
[4] INAF - Osservatorio Astronomico di Capodimonte, Saita Moiariello, 16, 80131 Napoli
[5] INAF - Fundacion Galileo Galilei,Santa Cruz de La Palma, Spain
[6] McDonald Observatory, The University of Texas at Austin, Austin, USA

Abstract

Since 1995, the extra-solar planet search has driven the high resolution spectroscopy community to build more and more stable spectrographs in order to reach the photon statistics limit in radial velocity measurements. This situation opened the possibility of asteroseismic observations of stellar p mode pulsations in solar-like stars. In this contribution we summarize the high precision radial velocity measurements of two solar type stars (α CMi and μ Her) using the SARG spectrograph at TNG equipped with an iodine cell. The analyzed spectra show individual measurement errors of about 1.0 m/s (very close to the theoretical photon noise limit). Further we discuss the synergy between high precision radial velocity asteroseismology campaigns and the search for Super Earths.

Individual Objects: Procyon, μ Her

Radial velocity uncertainties with SARG

SARG (Spettrografo Alta Risoluzione Galileo) is the high resolution optical spectrograph of the Italian Telescopio Nazionale Galileo (TNG). Instrument characteristics include a high spectral resolution (maximum about 160 000), high efficiency (peak at about 13%), rather large spectral coverage in a single shot. SARG was designed as a multipurpose instrument. The instrument, which mounts an iodine absorbing cell, is particularly suited for precise radial velocity programs, such as planet search and asteroseismology (Gratton et al. 2001). Radial velocities are obtained by means of the AUSTRAL code (Endl et al. 2000). We evaluate radial velocity uncertainties for SARG using the appropriate quality factor (see Bouchy et al. 2001) for spectrograph resolution, spectral type and rotational broadening. For stars brighter than 3.5 magnitude the resulting RV uncertainties range between 0.9 to 3 m/s depending on magnitude and spectral type of the star. This result affects also the efficiency of the instrument in detecting oscillations maintaining the number of spectra confined inside a reasonable number necessary to detect oscillation of amplitude ν_{osc} over a 4 σ threshold. We are aware that the limits of RV measurements are also set by spectrograph instabilities, but analyzing the results obtained with the subgiant star Procyon and μ Her we can limit the instrumental contribution to RV uncertainty lower than 1 m/s.

Procyon

Procyon A has been observed twice with SARG. The first time was in January 2001 (6 nights, 950 spectra) and the second was in January 2007 during a multisite campaign involving 11 Telescopes (Arentoft et al. 2008). In both campaigns a single the Doppler shift measurement has an rms internal error of 1.38 m/s. Sometimes, as shown by Claudi et al. (2005), the pulsation of the star is clearly visible in the RVs time series when there is no negative interference with other pulsation modes.

During the coordinated campaign, a coverage of Procyon lasting about 10 days without interruption was obtained. The resulting time series shows slow variations on a time scale of days. The good agreement between the measurements of the different telescopes lets one suspect that this is a variation that has a stellar origin, even if some contribution from instrumental drift could not be excluded. Nevertheless it could be interpreted as being due to rotational modulation from an active region on the stellar surface. The period is 10 days and could match the rotational period of this star or half of it (Arentoft et al. 2008). As far as the uncertainties of the RV measurements are concerned, HARPS is surely the best spectrograph. It is worth to mention that SARG has very similar RV uncertainties, as it is possible to see in Figure 1d of Arentoft et al. (2008).

μ Her

A clear detection of excess of power, providing a substantial evidence for solar-like oscillations in the G5 subgiant μ Her, was obtained with seven nights of observation with the SARG echelle spectrograph. The amplitude spectrum shows a clear excess of power centred at 1.2 mHz, with peak amplitudes of about 0.9 m/s. Successively, fitting the asymptotic relation to the power spectrum, a mode identification for the l=0,1,2,3 modes in the frequency range 900 − 1600 μHz is derived. The most likely value for the large separation turns out to be 56.5 μHz, consistent with theoretical expectations. The mean amplitude per mode (l=0 or 1) at peak power is 0.63 m/s, almost 3 times larger than the solar one. Results on this star are fully described by Bonanno et al. (2008).

Synergy with searches for Super Earths

The extension of the domain of radial velocity planet searches towards planets of masses significantly lower than Saturn is one of the most relevant results of the past years. The main difficulty is to strive for smaller radial velocity amplitude against the limits imposed by the star itself. It was shown that averaging consecutive radial velocity measurements it is possible to limit these effects (Santos et al. 2004).

In the search for solar-like oscillations targets are monitored for several contiguous nights. The data acquired are then very well suited to reach the limits of radial velocity precision.

To estimate the sensitivity of the our data to low-mass short-period planets we use the MULO code by M. Barbieri (described by Desidera et al. 2003). This allows us to constrain the masses of planets in circular as well as in eccentric orbits that are compatible or excluded by the data. Planets with masses larger than the derived limit are excluded because their presence would have created a detectable excess of variability with respect to the observed one. For both stars observed with SARG in asteroseismology campaigns (Procyon and μ Her) the nightly averages of the time series show an rms of about or less than 20 cm/s, that corresponds to limits on planetary masses in short period orbits of about 1.5 − 2.0 Earth masses. This result is based on only two targets (bright stars). Moreover each star has been observed in an optimal situation for what concerns the possible instrumental noise (only one target observed in successive nights for the whole run, keeping fixed the pointing and guiding). While we typically have a bit larger errors in normal planet-search survey operations (several targets per night over time scales of several years), nevertheless the result demonstrates that, at least on short temporal basis, a suitable observing strategy can push the intrinsic stellar limit of radial velocity measurements well below 1 m/s, allowing the detection of Earth-like planets in close orbits.

References

Arentoft, T., Kjeldsen, H., Bedding T. R., et al. 2008, ApJ, 687, 1180

Bonanno, A., Benatti, S., Claudi, R., et al. 2008, ApJ, 676, 1248

Bouchy, F., Pepe, F., & Queloz, D. 2001, A&A, 374, 733

Claudi, R., Bonanno A., Leccia S., et al. 2005, A&A, 429, L17

Desidera, S., Gratton, R., Endl, M., et al. 2003, A&A, 405, 207

Endl, M., Kürster, M., & Els, S. 2000, A&A, 374, 675

Gratton, R., Bonanno, G., Bruno, P., et al. 2001, Exp. Astr., 12, 107

Santos, N. C., Bouchy, F., Mayor, M., et al. 2004, A&A, 426, L19

Comm. in Asteroseismology,
Vol. 159, 2009, JENAM 2008 Symposium № 4: Asteroseismology and Stellar Evolution
S. Schuh & G. Handler

Using p-mode excitation rates for probing convection in solar-like stars

F. Kupka,[1] K. Belkacem,[2,3] M.-J. Goupil,[2] and R. Samadi[2]

[1] Max-Planck-Institute for Astrophysics, Karl-Schwarzschild Str. 1, 85748 Garching, Germany
[2] Observatoire de Paris, LESIA, CNRS UMR 8109, 92195 Meudon, France
[3] Institut d'Astrophysique et de Géophysique de l'Université de Liège, Allée du 6 Août 17, 4000 Liège, Belgium

Abstract

We discuss how the possibility to measure mode excitation rates through means of helio- and asteroseismology has improved our capabilities to test convection models and numerical simulations of surface convection and avoids ambiguities that have limited previous approaches.

Individual Objects: Sun, α Cen A

Probing Stellar Convection Models

Classical methods used in stellar evolution theory to "calibrate" the mixing length parameter α are based on integral quantities such as radius. However, if considered as an actual probe of convection models including mixing length theory itself such methods remain ambiguous (cf. also the study in Montalbán et al. 2004). Moreover, apart from the case of nearby stars, the basic stellar parameters required to derive these integral quantities can usually not be determined independently from the stellar structure models in which the convection models to be tested are actually used. A common alternative to this approach are tests which require to know physical conditions as a function of optical depth, for instance, the comparison of synthetic Balmer line profiles with observed ones (Gardiner at al. 1999, Heiter et al. 2002). A clear preference in favour of one particular convection model cannot be based on such comparisons either, except for the notion that none of the standard local convection models are able to permit predictions of detailed spectral line profiles for the entire range of A- to G-type stars. Furthermore, such tests have only limited implications on how reliably convection is modelled inside the star.

Seismology as an alternative?

That restriction does not hold for methods based on helio- and asteroseismology. Convection driven p-mode oscillations directly depend on the physical conditions of layers both near the stellar surface and inside the star. The comparison of predicted to observed p-mode frequencies has become a popular challenge for convection models used in stellar structure theory. Constraints such as the depth of the solar convection zone and the thickness of the region where deviations from an adiabatic temperature gradient occur ("overshooting") or the chemical composition (cf. Houdek 2009) probe convection models also in layers inside the star. Still, this approach can suffer from ambiguities when used to probe the critical superadiabatic layers of stellar convective envelopes, because the same acoustic size of a resonant cavity can originate from either a locally steeper temperature gradient (Basu & Antia 1995) or a more extended envelope due to turbulent pressure and the different efficiency of radiative cooling of a homogeneous (1D) compared to an inhomogeneous (3D) photosphere (Rosenthal et al. 1999). Hence, additional information is needed to remove these ambiguities and probe different properties of a convection model or a numerical simulation of convection.

Probing Convection with Excitation Rates

Fortunately, there is more information contained in helio- and asteroseismic observations: mode amplitudes and mode widths, which can now be obtained with direct measurements also from COROT (e.g., Appourchaux et al. 2008), and mode life times. Mode excitation due to shear stresses and entropy fluctuations is taken into account in a model by Samadi & Goupil (2001) which directly links the observed quantities with quantities such as the root mean square vertical velocity and convective flux as a function of depth (for an alternative approach and further references see also Houdek et al. 1999, for development of the physical concept see Gough 1980 and Christensen-Dalsgaard & Frandsen 1983, for a review Houdek 2006). The model requires eigenfunctions, associated eigenfrequencies, the mean structure, spatial and temporal correlations related to the turbulent kinetic energy and entropy, as well as the filling factor (relative horizontal area covered by regions of upflow) as a function of depth from a numerical simulation or a convection model. Seismological measurements and analyses provide the mode mass, the geometrical height at which modes are measured, the mode line width at half maximum, and the mean square of the mode surface velocity. Note that mode mass and geometrical height need some model input as well. With these data one can compute and compare predicted with observed mode excitation rates.

Results and further progress

Using this approach Samadi et al. (2006) were able to test several convection models widely used in the literature. Standard solar models (assuming a grey photosphere and a mixing length treatment of convection) underestimate the excitation of solar p modes by an order of magnitude. Samadi et al. (2006) could demonstrate that although both one-dimensional model atmospheres which account for non-grey radiative transfer and the full spectrum turbulence models as tested by Basu & Antia (1995) predict much higher mode excitation rates (up to a factor of 3), local convection models all fall clearly short of the observationally permitted error range. In earlier work Samadi et al. (2003) had shown that predictions derived from numerical simulations in turn do agree with observed data. Samadi et al. (2006) concluded that a non-local approach to convection modelling was needed which accounts for the asymmetric probability distributions of up- and down-, of hot and cold flows. A new model (Belkacem et al. 2006a) takes these properties into account. The "convection mode with plumes" is based on a model for higher order correlations of velocity and temperature for convective flows by Gryanik & Hartmann (2002). Though not all the input data for the model can yet be obtained without the use of numerical simulations, the model is a large step forward, since it was shown to reproduce p-mode excitation rates within observational uncertainties (Belkacem et al. 2006b). Recently, it was possible to repeat this work for the case of α Cen A (Samadi et al. 2008). The paper also analyzed the contributions of turbulent pressure, flow asymmetry, the time-correlation between eddies, and entropy fluctuations vs. Reynolds stresses. It was concluded that further a reduction of observational uncertainties (by typically a factor of 3) would allow asteroseismology to provide equally powerful tests as helioseismology now does for testing convection models.

Acknowledgments. F.K.'s work was possible thanks to a one month grant provided by the Observatoire de Paris. We are grateful to M.-A. Dupret and F. Baudin for their contributions to this research.

References

Appourchaux, T., Michel, E., Auvergne, M., et al. 2008, A&A, 488, 705

Basu, S., & Antia, H. M. 1995, ASP Conf. Ser., 76, 649

Belkacem, K., Samadi, R., Goupil, M.-J., & Kupka, F. 2006a, A&A, 460, 173

Belkacem, K., Samadi, R., Goupil, M.-J., et al. 2006b, A&A, 460, 183

Christensen-Dalsgaard, J., & Frandsen, S. 1983, Solar Physics, 82, 469

Gardiner, R. B., Kupka, F., & Smalley, B. 1999, A&A, 347, 876

Gough, D. O. 1980, in "Nonradial and Nonlinear Stellar Pulsation", eds. H. A. Hill and
     W. A. Dziembowski, Lecture Notes in Physics, 125, 273

Gryanik, V. M., & Hartmann, J. 2002, J. Atmos. Sci., 59, 2729

Heiter, U., Kupka, F., van 't Veer-Menneret, C., et al. 2002, A&A, 392, 619

Houdek, G. 2006, in "Beyond the spherical Sun", eds. K. Fletcher and M. Thompson, published on
     CDROM, ESA SP, 624, 28

Houdek, G., & Gough, D. O. 2009, CoAst, 159, 27

Houdek, G., Balmforth, N. J., Christensen-Dalsgaard, J., & Gough, D. O. 1999, A&A, 351, 582

Montalbán, J., D'Antona, F., Kupka, F., & Heiter, U. 2004, A&A, 416, 1081

Rosenthal, C. S., Christensen-Dalsgaard, J., Nordlund, Å, et al. 1999, A&A, 351, 689

Samadi, R., & Goupil, M.-J. 2001, A&A, 370, 136

Samadi, R., Nordlund, Å, Stein, R. F., et al. 2003, A&A, 404, 1129

Samadi, R., Kupka, F., Goupil, M.-J., et al. 2006, A&A, 445, 233

Samadi, R., Belkacem, K., Goupil, M.-J., et al. 2008, A&A, 489, 291

Comm. in Asteroseismology,
Vol. 159, 2009, JENAM 2008 Symposium № 4: Asteroseismology and Stellar Evolution
S. Schuh & G. Handler

Further progress on solar age calibration

G. Houdek [1] and D. O. Gough [1,2]

[1] Institute of Astronomy, Madingley Road, Cambridge CB3 0HA, UK
[2] Department of Applied Mathematics and Theoretical Physics, Cambridge CB3 0WA, UK

Abstract

We recalibrate a standard solar model seismologically to estimate the main-sequence age of the Sun. Our procedure differs from what we have done in the past by removing from the observed frequencies a crude representation of the effect of hydrogen ionization and the superadiabatic convective boundary layer. Our preliminary result is $t_\odot = 4.63 \pm 0.02$ Gy.

Individual Objects: Sun

Introduction

Seismological calibration of solar models to estimate the age of the Sun necessarily depends predominantly on the frequencies of the lowest-degree modes which penetrate into the energy-generating core where the greatest evolutionary change in the stratification occurs. Most commonly this is accomplished by fitting the asymptotic formula

$$\nu_{n,l} \sim \left[n + \frac{1}{2}l + \hat{\epsilon} - \sum_{k=1}^{K} \left(\sum_{j=0}^{k} A_{k,j} L^{2j} \right) \left(\frac{\nu_0}{\nu_{n,l}} \right)^{2k-1} \right] \nu_0 , \qquad (1)$$

to observed high-order frequencies $\nu_{on,l}$ of order n and degree l to determine the coefficients ν_0, $\hat{\epsilon}$, and $A_{k,j}$; here $L = l + \frac{1}{2}$. The most l−sensitive terms, at each degree $2k-1$, namely $A_{k,k}$, are, on the whole, the most sensitive to core conditions, and the least sensitive to the structure of the envelope (cf. Houdek & Gough 2007b). Therefore it is one or more of these that are the best determinants of stellar age. Eq. (1) is valid only if $l \ll n$ and n is large, such that the spatial scale of variation of the equilibrium state is everywhere much greater than the inverse vertical wavenumber of the mode. But that condition is not actually satisfied in the Sun: there is small-scale variation associated with ionization of abundant elements and the near discontinuity in low derivatives of the density at the base of the convection zone, which we call acoustic glitches, and which add the components $\nu_{gn,l}$ to $\nu_{n,l}$ that are in general oscillatory with respect to n. Ignoring these components introduces systematic errors into a straightforward fitting of Eq. (1) to $\nu_{on,l}$, errors that are evident in the undulatory age estimates as the limits of the frequency range adopted for the fitting are varied (Gough 2001). In an attempt to obviate these errors, Houdek & Gough (2007a, 2008) estimated the glitch components $\nu_{gn,l}$ by fitting to second differences (with respect to n) of the observed frequencies an asymptotic formula designed to represent the base of the convection zone and the two ionization zones of helium. In reality there is also an upper-glitch component, produced by the ionization of hydrogen and the upper superadiabatic boundary layer of the convection zone, which appears to be difficult to model in a reliable manner. When fitting the second differences Houdek &

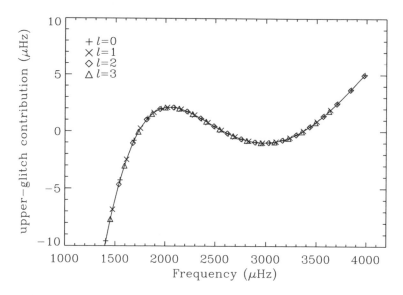

Figure 1: Contribution of the upper-glitch component to the glitch frequencies $\nu_{gn,l}$, obtained from the BiSON observations $\nu_{on,l}$ (Basu et al. 2007), as a function of $\nu_{on,l}$. It was obtained from summation of the series $P(\nu_{n,l})$, whose coefficients were determined from fitting to the second differences of the observed frequencies the second differences of an asymptotic formula representing the glitch components $\nu_{gn,l}$ (Houdek & Gough 2008). The upper-glitch component is produced by the ionization of hydrogen and the superadiabaticity of the surface boundary layer.

Gough (2007a) represented that component, coupled with the second differences of Eq. (1), somewhat arbitrarily as a series $P(\nu_{n,l})$ of inverse powers of $\nu_{n,l}$. Because the upper-glitch component is relatively smooth, they subsequently tacitly regarded it as being included in the smooth asymptotic expression (1) by adjusting the observed frequencies by only the component $\nu_{gn,l}$. Because the upper glitch is quite close to the surface (partly in the evanescent zones of the modes), its influence on the eigenfrequencies is essentially independent of l, and so should not have materially affected the fitted coefficients $A_{k,j}$ with $j > 0$.

Modification to the calibration procedure

In the work we report here we have tested the stability of the procedure by including in $\nu_{gn,l}$ a representation of the upper-glitch component. To this end we summed the second-difference representation P to obtain an estimate of its contribution to the frequencies. There is some ambiguity in how one separates smooth and glitch components near the surface, which is exhibited by the two undetermined constants of summation of the second differences; here we chose those constants by minimizing the error-weighted sum of the squares of the upper-glitch frequencies. The outcome is plotted in Fig. 1 using BiSON data (e.g. Basu et al. 2007) up to degree $l = 3$.

After fitting Eq. (1), with $K = 3$, to the resulting glitch-adjusted observed frequencies, the coefficients ν_0, $\hat{\epsilon}$ and $A_{k,0}$ were found to be naturally somewhat different from the results obtained without the upper-glitch adjustment. But the coefficients $A_{k,k}$ are similar. There is, however, a slight difference, which is evidently a product of an inadequacy of the asymptotic formulae to reproduce precisely the observed frequencies of the Sun.

Result

The result of the present model calibration against BiSON data (e.g. Basu et al. 2007) is

$$t_\odot = 4.63 \pm 0.02 \, \text{Gy} , \qquad (2)$$

a value in fair accord with our previous estimates (Houdek & Gough 2007b, 2008). The errors quoted here come solely from the stated observed frequency errors, which we have assumed to be statistically independent, and take no account of (systematic) errors in our procedure; that the value (2) differs from our previous estimates by as much as $2.5\,\sigma$ suggests that such systematic errors could be present at a level at least as great as the random errors. Our current value for the solar age is lower than the previous estimates by essentially this method, although, in contrast to many earlier estimates, it remains greater than the age of Model S of Christensen-Dalsgaard et al. (1996), which we used as our reference. It is also greater than that of many, if not all, meteorites. We have not yet completed our investigation of the robustness of the result, so we offer it still as a preliminary estimate.

Acknowledgments.   GH is grateful for support by the Science and Technology Facilities Council.

References

Basu, S., Chaplin, W. J., Elsworth, Y., et al. 2007, ApJ, 655, 660

Christensen-Dalsgaard, J., Däppen., W., Ajukov, S. V., et al. 1996, Science, 272, 1286

Gough, D. O. 2001, in "Astrophysical Ages and Timescales", eds. T. von Hippel, C. Simpson, and N. Manset, ASP Conf. Ser., 245, 31

Houdek, G., & Gough, D. O. 2007a, MNRAS, 375, 861

Houdek, G., & Gough, D. O. 2007b, in "Unsolved Problems in Stellar Physics", eds. R. J. Stancliffe, J. Dewi, G. Houdek, et al., AIP Conf. Proc., 948, 219

Houdek, G., & Gough, D. O. 2008, in "The Art of Modelling Stars in the 21st Century", eds. L. Deng, K. L. Chan, and C. Chiosi, IAU Symposium, 252, 149

Comm. in Asteroseismology,
Vol. 159, 2009, JENAM 2008 Symposium № 4: Asteroseismology and Stellar Evolution
S. Schuh & G. Handler

Parametric interaction of coronal loops with p modes

A. V. Stepanov,[1] V. V. Zaitsev,[2] A. G. Kisliakov,[2] and S. Urpo[3]

[1] Pulkovo Observatory, Pulkovo chaussee 65, 196140 Saint Petersburg, Russia
[2] Institute of Applied Physics, Ulianova str. 46, 603950 Nizhny Novgorod, Russia
[3] Metsähovi Radio Observatory, Metsähovintie 114, 02540 Kylmälä, Finland

Abstract

Parametric resonance between p modes and eigenoscillations of coronal loops is studied. Observations of solar radio bursts revealed this effect in simultaneous excitation of loop oscillations with periods corresponding to the pumping-up frequency (5 min), subharmonic (10 min), and to the first upper frequency of parametric resonance (3.3 min). An interpretation in terms of a coronal magnetic loop as an equivalent electric circuit is given. Parametric resonance can work as a channel for transfer of energy from photospheric motions to stellar coronae.

Individual Objects: Sun

Introduction

The heating of solar (up to 10^6–10^7 K) and stellar coronae (10^7–10^8 K) has been explained by a variety of mechanisms. In particular acoustic heating, reconnection, damping of MHD-waves, topological dissipation, microflares were discussed in this context, but without an obvious solution in sight. We consider here another possible origin of coronal heating dealing with parametric resonance between coronal loops and p modes.

Parametric excitation of loop oscillations by p modes

There is quite a lot of evidence of 5-min oscillations in the solar corona as inferred from TRACE and SOHO data (see e.g. De Moortel et al. 2002; De Pontieu et al. 2005) and Metsähovi observations at 11 and 37 GHz (Kislyakov et al. 2006; Zaitsev & Kislyakov 2006). Figure 1 presents the example of spectral analysis of the solar flare event on March 20, 2000 which revealed 10, 5, and 3.3 min oscillations. Why are p-mode oscillations driven by photosphere convection observed at the coronal level?

De Pontieu et al. (2005) suggested a tunneling or direct propagation of photospheric oscillations into the outer atmosphere via slow MHD and shock waves. We interpret such kinds of events using the electric circuit model for a coronal loop (Zaitsev et al. 2000) and parametric resonance between acoustic eigenmodes of a loop and p modes. Indeed, 5-min velocity oscillations of the photosphere modulate the electric current in a coronal loop foot point. Parametric resonance between photosphere oscillations and sound oscillations $\omega_0 = k_{||} c_s$ of the loop appears. This effect reveals itself in the excitation of loop oscillations with periods of 5 min (pumping-up frequency ω), 10 min (sub-harmonic $\omega/2$), and 3.3 min

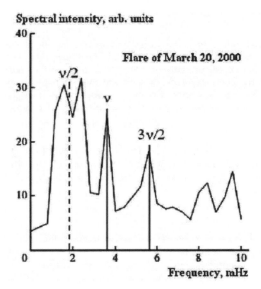

Figure 1: Spectrum of pulsations in the flare event on March 20, 2000 observed with the Metsähovi radio telescope at 37 GHz. The dashed line shows the frequency of the 1.8 mHz (10 min period), while the solid vertical lines indicate the frequencies of 5-min (3.6 mHz) and 3.3-min (5.6 mHz) oscillations.

(corresponding to the first upper frequency of parametric resonance $3\omega/2$). Modulation of electric current leads to modulation of sound velocity in a loop (Zaitsev & Kislyakov 2006):

$$c_s = \left(\frac{2\gamma k_B T}{m_i}\right)^{1/2}\left(1 + \frac{\Delta T}{2T}\right) = c_{s0}(1 + 0.5q\cos\omega t), \qquad q = \frac{4}{3}\frac{\gamma - 1}{\gamma}\frac{I^2}{\pi c^2 r^2 p}\frac{I_\sim}{I}.$$

Here p is the gas pressure, r is the loop radius, $k_\| = s\pi/L$, L is the loop length, $s = 1, 2, 3,$ $\ldots$, $I \gg I_\sim$. Deviations of loop plasma parameters can be written as

$$\frac{d^2 y}{dt^2} + \omega_0^2(1 + q\cos\omega t)y = 0$$

Here the parameter q is determined by the width of zones near the frequencies of a parametric resonance $\omega_n = n\omega/2$, $n = 1, 2, 3, \ldots$ where the parametric instability occurs (Landau & Lifshitz 1976). The width of the first instability zone $(-q\omega_0/2 < \omega/2 - \omega_0 < q\omega_0/2)$, where ω_0 is close to $\omega/2$, is larger. So, the amplitude of variations of sound speed under action of photospheric oscillations is higher compared to the second considerably narrowed zone $\omega \approx \omega_0$ if $q \ll 1$ (Fig. 1). Since p modes cannot directly penetrate into the corona, the parametric resonance may work as an effective channel for transfer of energy of photospheric oscillations to the upper layers of the atmosphere. It offers the challenge to understanding the heating mechanism of stellar coronae. Estimations made for solar loops have shown that if the current exceeds 7×10^9 A, the energy flux of sound oscillations arising in the coronal magnetic loop as a result of parametric resonance exceeds losses due to optical radiation.

Conclusions

Hindman & Jain (2008) proposed recently that loop oscillations are generated due to the buffeting excitation of loop magnetic fibrils by p modes. They supposed however that magnetic fields are potential and no electric current exists. Our mechanism suggests the interaction of p modes with electric currents in coronal loop and the heating of loop foot points due to the dissipation of the electric currents. Because p modes do not penetrate into the corona, the parametric resonance is a good way for transfer of energy of photosphere oscillations to stellar coronae.

Acknowledgments.   This work was partially supported by RFBR grant 09-02-00624-a, the Program of Russian Academy of Sciences "Solar Activity and Physical Processes in the Sun-Earth System", and the grants of Leading Scientific Schools NSH-4485.2008.2 and NSH-6110.2008.2.

References

De Moortel, I., Ireland, J., Hood, A. W., & Walsh, R. W. 2002, A&A, 387, L13

De Pontieu, D., Erdelyi, R., & De Moortel, I. 2005, ApJ, 624, L61

Hindman, B. W., & Jain, R. 2008, ApJ, 677, 769

Kislyakov, A. G., Zaitsev, V. V., Stepanov, A. V., & Urpo, S. 2006, Solar Phys., 233, 89

Landau, L. D., & Lifshitz, E. M. 1976, in "Mechanics", Pergamon Press, Oxford

Zaitsev, V. V., & Kislyakov, A. G. 2006, Astron. Rep., 50, 823

Zaitsev, V. V., Urpo, S., & Stepanov, A. V. 2000, A&A, 357, 1105

Reception at the University Observatory.

Comm. in Asteroseismology,
Vol. 159, 2009, JENAM 2008 Symposium № 4: Asteroseismology and Stellar Evolution
S. Schuh & G. Handler

A preliminary glimpse on CoRoT results and expectations

E. Michel,[1] K. Zwintz,[2] and the CoRoT team

[1] Observatoire de Paris, LESIA, UMR 8109, pl. J. Janssen, 92195 Meudon, France
[2] Institut für Astronomie, Universität Wien, Türkenschanzstraße 17, 1180 Vienna, Austria

Abstract

The full content of this review paper has been presented by Michel et al. (2008).

References

Michel, E., Baglin, A., Weiss, W. W., et al. 2008, CoAst, 156, 73

Comm. in Asteroseismology,
Vol. 159, 2009, JENAM 2008 Symposium № 4: Asteroseismology and Stellar Evolution
S. Schuh & G. Handler

Asteroseismology of B stars

A. Thoul [1,2]

[1] Université de Liège, 17 allée du 6 août, Bât B5c, B-4000 Liège, Belgium

Abstract

Numerous B stars on the main sequence are found to be variable. These stars have a relatively simple structure, and yet they present some very interesting challenges. It is important to understand these stars before we can expect to be able to understand stars which have a more complicated internal structure.

Asteroseismology of B stars has made considerable progress in the last few years, thanks to the observations obtained from large multisite campaigns, and it is now possible to determine global parameters for these stars such as their masses, ages, metallicities, with very high accuracy. Many variable B stars are also observed with the CoRoT space mission, and statistical studies may become possible in the near future.

Detailed studies of the oscillation spectra of β Cephei stars have already allowed to put some limits on the overshooting parameter, and this overshooting parameter is found to vary from one star to the next. Some β Cephei stars are found to present differential rotation in their envelopes, while others are compatible with solid body rotation. Some B stars present nitrogen enhancement, even though they are very slow rotators. The instability strips of B stars differ depending on the composition and the opacity tables adopted. Microscopic diffusion and radiative accelerations could produce an accumulation of iron-group elements in some layers of these stars. Hybrid pulsators, showing both β Cephei and SPB pulsations have been observed. Some of the best observed stars have pulsation spectra which still cannot be reproduced by modelling.

Here I review the present status of the observations and of the modelling of β Cephei stars, emphasizing both the successes reached and the questions and problems which remain open today.

Individual Objects: 16 Lac, HD 129929, ν Eri, θ Oph, 12 Lac

Introduction

Classical stellar observables, such as the effective temperature, the surface gravity, the luminosity, and the metallicity, are not sensitive to the internal structure of the stars, and by themselves do not give enough information for the determination of the basic stellar parameters. However, pulsation frequencies are very sensitive to the details of the internal structure. By comparing the oscillation spectra to those computed from theoretical models, it is possible to obtain accurate global stellar parameters, such as stellar masses, radii, temperatures, luminosities, metallicities, but also to improve the physics used in the theoretical models, such as the physics of convection, rotational processes, mixing, equation of state, nuclear reaction rates, opacities, microscopic diffusion, radiative accelerations, mass loss, and magnetism.

---

[2] Chercheur Qualifié FNRS

Pulsating main sequence B stars can be classified as β Cephei, SPB or Be stars. They have a relatively simple internal structure, with a massive convective core and a radiative envelope. The pulsations in these stars are excited by the κ mechanism due to the iron opacity bump at 200 000 K. These are long-lived intrinsically unstable modes. Long-term multisite photometric and spectroscopic campaigns are needed to resolve the oscillation frequencies of these stars. Several such campaigns have been organized to observe single β Cephei stars; detailed asteroseismic studies of these stars have been performed, and the results are described in the next section. Both photometric and spectroscopic observations are needed because they sample the same pulsations, but with different sensitivity to the degree and azimuthal order of the modes. Photometric observations measure weighted averages of the pulsation amplitude over the $\tau = 2/3$ surface, so they are mostly sensitive to low-degree modes. Doppler observations measure the projection of the velocity over the line of sight, and are therefore slightly more sensitive to modes of moderate degree l. Photometry is commonly used to observe pulsations in main sequence B stars. Multicolour photometry and spectroscopy are however essential, because, as we will show later, it is necessary to identify the modes to do useful asteroseismology with these stars. The amplitude and phase behaviour of an oscillation mode are different in different filters, so that the degree l can be determined through multi-colour photometry. Line-profile variations are determined by parameters including the degrees l and azimuthal orders m of all the pulsational velocities.

Pulsating Be stars are fast rotators with emission lines. Asteroseismology of these complicated stars is still in its infancy. SPB stars have masses between 4 and 7 $M_\odot$. They pulsate in the asymptotic regime of high-order g modes with periods of 0.5 to 5 days. This is a serious observational difficulty, and very long multisite campaigns are needed. At this time, much fewer modes are detected than expected, and they cannot be identified due to their low amplitudes. The asteroseismology of SPB stars is therefore also still in its infancy. β Cephei stars are more massive; their masses range between 8 and 18 $M_\odot$. They are mostly slow rotators, and have a rather sparse spectrum of low-degree, low-order p, g and mixed modes, with periods of 2 to 8 hours. In β Cephei stars, the individual frequencies probe different layers of the star, and therefore, each frequency in the spectrum gives independent information about the stellar interior.

Detailed asteroseismic studies of β Cephei stars

Asteroseismic studies of β Cephei stars consist of performing direct frequency fitting of axisymmetric modes. This can provide accurate determinations of basic stellar parameters, constraints on the core overshooting parameters and on the metallicity, or it can also lead to inconsistencies, i.e., information on missing or incorrect physics in the stellar models. The multiplets are then analyzed, and the splittings are used to get information on the rotation rate in the stellar envelope.

Main-sequence B stars are modelled using "standard" stellar evolution codes. The modelling parameters are typically the stellar mass M, the initial hydrogen mass fraction X, the metallicity Z, and the overshooting parameter α_{ov}, which is a measure of the extent of the mixed region at the boundary of the convective core. Several different prescriptions can be used for some of the physics needed in the stellar models, such as the theory of convection, opacity tables, gravitational settling and radiative accelerations, the initial chemical abundances, rotation ...

Here we give the main results which were obtained from detailed modelling of several β Cephei stars, the successes and problems encountered, and conclusions that were reached.

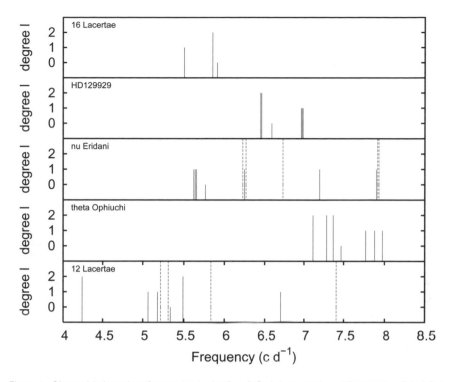

Figure 1: Observed independent frequencies in the five β Cephei stars 16 Lac, HD 129929, ν Eri, θ Oph and 12 Lac. Solid lines are used for identified modes, while dashed lines are used for modes with no definite identification.

16 Lacertae

16 Lacertae is a spectroscopic binary and eclipsing variable. Its luminosity and effective temperature are constrained well. Three oscillation frequencies have been detected in this β Cephei star. The degrees of these three modes of oscillation have been identified through spectroscopic and photometric mode identification methods (Chapellier et al. 1995, Dziembowski & Jerzykiewicz 1996, Aerts et al. 2003, Thoul et al. 2003). Two of the modes are axisymmetric, but there are no multiplets. The frequencies of the three main modes of pulsation observed in 16 Lac are $f_1 = 5.9112\,\mathrm{cd}^{-1}$, which is a radial mode, $f_2 = 5.8551\,\mathrm{cd}^{-1}$, which is an $l = 2$ axisymmetric mode, and $f_3 = 5.5033\,\mathrm{cd}^{-1}$, which is an $l = 1$ mode. They are represented in Fig. 1.

Evolutionary sequences have been calculated with the stellar evolution code CLES (Scuflaire et al. 2008a) and oscillation frequencies for each model were calculated using the adiabatic stellar oscillations code OSCL (Scuflaire et al. 2008b). For each set of model parameters (M, X, Z, α_{ov}), fitting the radial mode gives the age of the star, or, equivalently, its effective temperature T_{eff}. For a given set of model parameters (X, Z, α_{ov}), fitting the second axisymmetric frequency fixes the age and the mass M. It is interesting to note that T_{eff} and $\log g$ are not imposed in the models; their values result from the models which fit the observed frequencies. The observed error box is however extremely useful to constrain the possible ranges for the stellar parameters.

The mode with frequency f_1 was identified by the modelling as an $l = 0$ p_1 mode, f_2 as an $l = 2$ g_1 mode and f_3 as an $l = 1$ g_1 mode; this last one is in fact a mixed mode. The models were performed for only two values of α_{ov}, namely 0 and 0.2. Very precise values were obtained for the stellar mass, metallicity, effective temperature, luminosity and age (Thoul et al. 2003).

HD 129929

The star HD 129929 has been observed for over 20 years. Six independent pulsation frequencies were observed and identified, including one radial mode, one complete triplet and two members of a quintuplet (Aerts et al. 2004). Detailed asteroseismology of this star is possible since two axisymmetric modes are observed, and information about the internal rotation rate can be obtained from the two multiplets (Aerts et al. 2003b, Dupret et al. 2004).

The frequencies of the six independent modes of pulsation observed in HD 129929 are $f_1 = 6.461699\,\mathrm{cd}^{-1}$, $f_2 = 6.978305\,\mathrm{cd}^{-1}$, $f_3 = 6.449590\,\mathrm{cd}^{-1}$, $f_4 = 6.990431\,\mathrm{cd}^{-1}$, $f_5 = 6.590940\,\mathrm{cd}^{-1}$, and $f_6 = 6.966172\,\mathrm{cd}^{-1}$. f_5 is a radial mode, f_2, f_4 and f_6 form an $l=1$ triplet, and f_1 and f_3 are two members of a quintuplet. They are represented in Fig. 1.

Seismic modelling of HD 129929 was performed using CLES and OSCL. The radial mode is identified by the modelling as a p_1 mode, the central frequency of the triplet corresponds to a p_1 mode, and f_1 and f_3 are confirmed to be members of an $l = 2$ g_1 quintuplet. The excitation of the modes was analyzed using MAD, a non-adiabatic oscillation code (Dupret et al. 2003). Constraints on the metallicity were obtained by requiring good compatibility between the observed and calculated amplitude ratios, and that the observed modes must be excited; this excluded solutions with no overshooting. In addition, in order to find solutions that fit the quintuplet, solutions with an overshooting larger than 0.15 had to be excluded also. Very precise values were obtained for the effective temperature, luminosity, mass, and age of the star.

The rotational splitting integrals of the two multiplets have amplitude peaks at different depths in the star; the two multiplets therefore give information about the rotation rates at different layers inside the star. It was possible to rule out rigid rotation in the star's envelope (Aerts et al. 2003b, Dupret et al. 2004).

ν Eridani

Two large multisite photometric campaigns and one large multisite spectroscopic campaign were dedicated to the star ν Eridani. From these campaigns, twelve independent frequencies were obtained and seven of these frequencies were identified (Handler et al. 2004, Aerts et al. 2004b, De Ridder et al. 2004, Jerzykiewicz et al. 2005). The frequencies of ν Eri are $f_1 = 5.7632828\,\mathrm{cd}^{-1}$, $f_2 = 5.6538767\,\mathrm{cd}^{-1}$, $f_3 = 5.6200186\,\mathrm{cd}^{-1}$, $f_4 = 5.6372470\,\mathrm{cd}^{-1}$, $f_5 = 7.898200\,\mathrm{cd}^{-1}$, $f_6 = 6.243847\,\mathrm{cd}^{-1}$, $f_7 = 6.262917\,\mathrm{cd}^{-1}$, $f_8 = 7.20090\,\mathrm{cd}^{-1}$, $f_9 = 7.91383\,\mathrm{cd}^{-1}$, $f_{10} = 7.92992\,\mathrm{cd}^{-1}$, $f_{11} = 6.73223\,\mathrm{cd}^{-1}$, $f_{12} = 6.22360\,\mathrm{cd}^{-1}$. In addition, two low-frequency modes typical of SPB pulsations were observed, with frequencies $f_A = 0.432786\,\mathrm{cd}^{-1}$ and $f_B = 0.61440\,\mathrm{cd}^{-1}$. ν Eri has been called a "hybrid β Cephei/SPB pulsator". The frequency f_1 is identified as a radial mode, the frequencies f_2, f_3 and f_4 form an $l = 1$ triplet; f_5, f_6, f_8 are identified as $l = 1$ modes. They are shown in Fig. 1.

Detailed modelling of this star has been performed by several groups (Ausseloos et al. 2004, Pamyatnykh et al. 2004, Dziembowski & Pamyatnykh 2008). Solutions were found that reproduced the observed frequencies of ν Eri, but those frequencies were not excited. Some authors suggested that the modes could be excited if iron accumulates in the excitation region due to microscopic diffusion. Others suggested that non-standard stellar models should be considered. In both cases, some of the high-frequency modes were excited and the two low-frequency modes were still stable. Recently, changes in solar abundances were reported

(Asplund et al. 2005), and new opacity tables made available (Badnell et al. 2005). It was shown that these new abundances and opacities displace the blue edge of the instability strips of β Cephei and SPB stars towards higher temperatures and increase considerably the potential number of hybrid pulsators (Miglio et al. 2007). Using these new abundances and opacity tables in the modelling of ν Eri, solutions were found where all high frequency modes (except for the highest one) are excited, and where a range of g modes is excited, which includes one of the two low-frequency modes observed (Dziembowski & Pamyatnykh 2008).

Modelling of ν Eri was performed using the stellar evolution code CLES (Scuflaire et al. 2008a) with the Asplund et al. (2005) abundances and the new OP opacities (Badnell et al. 2005), and the oscillation frequencies were calculated using the oscillation code OSCL (Scuflaire et al. 2008b), for two values of the hydrogen mass fraction, namely $X = 0.72$ and $X = 0.70$. Models are found which fit exactly the four axisymmetric modes: f_1 is a radial p_1 mode, f_4 is an $l = 1$ g_1 mode, f_6 is an $l = 1$ p_1 mode, and f_8 is an $l = 1$ p_2 mode. The mass of the star is $M = 9.0 \pm 0.1\,M_\odot$, the overshooting parameter is $\alpha_{ov} = 0.22$, and the metallicity $Z = 0.021 \pm 0.001$. The high frequency modes f_8, f_5, f_9 and f_{10} are not excited. Low-frequency high-order g modes are excited in the range $0.55 - 0.91\,\mathrm{cd}^{-1}$. In these models, f_B is excited, but not f_A.

The problems encountered in modelling the β Cephei star ν Eri have been mostly solved when the new solar abundance and the new OP opacity tables were used. Some problems remain however, since the highest frequency observed and one of the low frequency modes are still not excited.

θ Ophiuchi

Photometric and spectroscopic campaigns were dedicated to the β Cephei star θ Ophiuchi (Handler et al. 2005, Briquet et al. 2005). Seven independent frequencies were observed in a narrow frequency range; they were all identified, and included one radial mode, $f_3 = 7.4677\,\mathrm{cd}^{-1}$, one triplet, $f_4 = 7.7659\,\mathrm{cd}^{-1}$, $f_6 = 7.8742\,\mathrm{cd}^{-1}$, $f_7 = 7.9734\,\mathrm{cd}^{-1}$; and three members of a quintuplet, $f_1 = 7.11600\,\mathrm{cd}^{-1}$, $f_5 = 7.2881\,\mathrm{cd}^{-1}$, $f_2 = 7.3697\,\mathrm{cd}^{-1}$. They are represented in Fig. 1.

Detailed seismic modelling was performed for this β Cephei star using the codes CLES and OSCL, with the new solar abundances, the increased abundance of neon from Cunha et al. (2006), and the new OP opacities, for two values of the hydrogen mass fraction, namely $X = 0.71$ and $X = 0.72$, and for metallicities $Z = [0.009, 0.015]$. Models are chosen such that the two axisymmetric modes are reproduced; the quintuplet is used to obtain additional constraints. The models have a high overshooting parameter, $\alpha_{ov} = 0.44 \pm 0.07$ and a mass $M = 8.2 \pm 0.3\,M_\odot$. All solutions are inside the photometric error box, but outside the 2σ spectroscopic error box. All observed modes are excited. The unequal splittings of the triplet can be reproduced by taking into account second order effects of rotation (Briquet et al. 2007).

12 Lacertae

12 Lacertae is probably the most puzzling β Cephei star observed so far. It has been full of surprises. It had already been observed as a variable star in 1915 (Young 1915), with frequency $1/P = 5.17896\,\mathrm{cd}^{-1}$, a value very close to the present value of its main mode of oscillation, $5.179034\,\mathrm{cd}^{-1}$. It was already observed as a multiperiodic variable star in 1957 (Abrami 1957). Two frequencies of oscillation have been known with great accuracy since 1961 (Opalski & Ciurla 1961): $f_1 = 5.17897\,\mathrm{cd}^{-1}$ and $f_2 = 5.06666\,\mathrm{cd}^{-1}$. The present values for these two frequencies are $5.179034\,\mathrm{cd}^{-1}$ and $5.066346\,\mathrm{cd}^{-1}$. In 1978, six frequencies were determined, including an $l = 3$ equidistant triplet (Jerzykiewicz 1978, Jarzebowski et al. 1980). The modes of the triplet were later contrarily identified as $l = 2$ (Smith

1980). Attempts were made in 1994 to discriminate between the different identifications using the moment method, without success (Mathias 1994). Identification was attempted through modelling in 1999 (Dziembowski & Jerzykiewicz 1999), again with no result. In fact modelling of 12 Lac seems to be problematic. In 2006, a large multisite campaign was dedicated to 12 Lac. Ten independent frequencies were observed, six were identified, with a big surprise: The three frequencies in the very equidistant triplet are identified with three different values of the degree l: they do not form a triplet after all (Handler et al. 2006)! The frequencies observed are $f_1 = 5.179034\,\mathrm{cd}^{-1}$, $f_2 = 5.066346\,\mathrm{cd}^{-1}$, which are both identified as $l = 1$ modes, $f_3 = 5.490167\,\mathrm{cd}^{-1}$, identified as an $l = 2$ mode, $f_4 = 5.334357\,\mathrm{cd}^{-1}$, a radial mode, $f_5 = 4.24062\,\mathrm{cd}^{-1}$, identified as $l = 2$, $f_6 = 7.40705\,\mathrm{cd}^{-1}$, $f_7 = 5.30912\,\mathrm{cd}^{-1}$, $f_8 = 5.2162\,\mathrm{cd}^{-1}$, $f_9 = 6.7023\,\mathrm{cd}^{-1}$, identified as an $l = 1$ mode, $f_{10} = 5.8341\,\mathrm{cd}^{-1}$, and a low-frequency mode of frequency $f_A = 0.35529\,\mathrm{cd}^{-1}$ (Desmet et al. 2007). These frequencies are represented in Fig. 1, except for the low-frequency g mode.

A first attempt at modelling 12 Lac with these new identifications was performed by Dziembowski & Pamyatnykh (2008). They assumed a hydrogen mass fraction of $X = 0.70$, a metallicity $Z = 0.015$, a chemical abundance determined by the solar composition of Asplund et al. (2004), and they used the new OP opacities. They used the following error boxes for the position in the HR diagram: $\log T_{\mathrm{eff}} = [4.355, 4.395]$, $\log L = [4.0, 4.35]$. Their calculations were performed without overshooting. They fitted the mode corresponding to f_4 with the radial fundamental mode, and f_2 with an $l = 1$ g_1 mode. They excluded any other possible identification. Desmet et al. (2009) recently performed a detailed modelling of 12 Lac, using CLES with the newest solar abundances (Asplund et al. 2005), OP opacities, $X = 0.72$, $Z = 0.015$, and an error box in the HR diagram given by $\log T_{\mathrm{eff}} = 4.389 \pm 0.018$, $\log g = 3.65 \pm 0.15$. Using the identification of f_9, they rule out f_4 as the radial fundamental, and identify this mode as the first overtone. The range of theoretically predicted unstable modes does not cover the whole range of observed modes; the two highest frequencies are not excited, and the SPB-type pulsation mode f_A is not excited.

Conclusions

Detailed asteroseismic studies of a few β Cephei stars have been very successful. They have shown that very precise determinations of basic stellar parameters such as the mass, luminosity, effective temperature, metallicity, and age can be obtained. In addition, precise values of the overshooting parameter can be derived. So far, α_{ov} takes values from 0 to 0.4 and is found to vary from one star to another. In some stars, zero overshooting can be ruled out. Evidence for non-rigid rotation in the stellar envelope has been found for at least one β Cephei star, namely HD 129929. The modelling of ν Eri has been problematic. Using the new solar abundances and the new OP opacity tables in the models have helped to reduce considerably the discrepancy between the observations and the theoretical models, but the problem is not completely solved. The long and complicated story of the mode identification of the frequencies observed in 12 Lacertae has shown that good and reliable mode identification, using both spectroscopic observations and multicolour photometry, is essential before any useful detailed modelling can be performed.

The observational challenges are therefore to obtain very good frequency determinations, which is particularly difficult for SPB stars due to the long periods of the high-order g modes, and reliable mode identifications, which are sometimes very tricky since multiplets due to rotational splittings can overlap with the spectrum of axisymmetric frequencies. A good determination of basic stellar parameters (T_{eff}, $\log g$, $[M/H]$) is very helpful to rule out some of the models and reduce the parameter space. Therefore, even though a huge amount of photometric data will soon be available from space observations, it is imperative to perform ground-based follow-up observations with multicolour photometry and high-resolution spectroscopy.

The main theoretical challenges at this time are to explain the range of excited modes in β Cephei stars, the presence of β Cephei and SPB stars in low metallicity environments, and the existence of hybrid SPB/β Cephei pulsators. Asteroseismic studies of B type pulsators will give constraints which will lead to improvements in the physics used in the models, such as the modelling of rotation, mixing, convection, ... As the study of ν Eri has proven, reliable opacity tables as well as correct initial chemical compositions are crucial.

Acknowledgments. AT is grateful to the organizers of the Symposium for inviting her to give a review talk on this topic, and to HELAS for providing financial support.

References

Abrami, A. 1957, Nature, 180, 1112

Aerts, C., Lehmann, H., Briquet, M., et al. 2003, A&A, 399, 639

Aerts, C., Thoul, A., Daszynska, J., et al. 2003b, Science, 5627, 1926

Aerts, C., Waelkens, C., Daszynska-Daszkiewicz, J., et al. 2004, A&A, 415, 241

Aerts, C., De Cat, P., Handler, G., et al. 2004b, MNRAS, 347, 463

Asplund, M., Grevesse, N., Sauval, A. J., et al. 2004, A&A, 417, 751

Asplund, M., Grevesse, N., Sauval, A. J., et al. 2005, A&A, 431, 693

Ausseloos, M., Scuflaire, R., Thoul, A., & Aerts, C. 2004, MNRAS, 355, 352

Badnell, N. R., Bautista, M. A., Butler, K., et al. 2005, MNRAS, 360, 458

Briquet, M., Lefever, K., Uytterhoeven, K., & Aerts, C. 2005, MNRAS, 362, 619

Briquet, M., Morel, T., Thoul, A., et al. 2007, MNRAS, 381, 1482

Chapellier, E., Le Contel, J. M., Le Contel, D., et al. 1995, A&A, 304, 406

Cunha, K., Hubeny, I., & Lanz, T. 2006, ApJ, 647, L143

De Ridder, J., Telting, J. H., Balona, L. A., et al. 2004, MNRAS, 351, 324

Desmet, M., Briquet, M., De Cat, P., et al. 2007, CoAst, 150, 195

Desmet, M., Briquet, M., Thoul, A., et al. 2009, MNRAS, submitted

Dupret, M.-A., De Ridder, J., De Cat, P., et al. 2003, A&A, 398, 677

Dupret, M.-A., Thoul, A., Scuflaire, R., et al. 2004, A&A, 415, 251

Dziembowski, W. A., & Jerzykiewicz, M. 1996, A&A, 306, 436

Dziembowski, W. A., & Jerzykiewicz, M. 1999, A&A, 341, 480

Dziembowski, W. A., & Pamyatnykh, A. A. 2008, MNRAS, 385, 2061

Handler, G., Shobbrook, R. R., Jerzykiewicz, M., et al. 2004, MNRAS, 347, 454

Handler, G., Shobbrook, R. R., & Mokgwetsi, T. 2005, MNRAS, 362, 612

Handler, G., Jerzykiewicz, M., Rodriguez, E., et al. 2006, MNRAS, 365, 327

Jarzebowski, T., Jerzykiewicz, M., Rios Herrera, M., & Rios Berumen, M. 1980, RMxAA, 5, 31

Jerzykiewicz, M. 1978, Acta Astronomica, 28, 465

Jerzykiewicz, M., Handler, G., Shobbrook, R. R., et al. 2005, MNRAS, 360, 619

Mathias, P., Aerts, C., Gillet, D., & Waelkens, C. 1994, A&A, 289, 875

Miglio, A., Montalbán, J., & Dupret, M.-A. 2007, MNRAS, 375, L21

Opolski, A., & Ciurla, T. 1961, Acta Astronomica, 11, 231

Pamyatnykh, A. A., Handler, G., & Dziembowski, W. A. 2004, MNRAS, 350, 1022

Scuflaire, R., Théado, S., Montalbán, J., et al. 2008a, Ap&SS, 316, 83

Scuflaire, R., Montalbán, J., Théado, S., et al. 2008b, Ap&SS, 316, 149

Smith, M. A. 1980, ApJ, 240, 149

Thoul, A., Aerts, C., Dupret, M. A., et al. 2003, A&A, 406, 287

Young, R. K. 1915, JRASC, 9, 423

Comm. in Asteroseismology,
Vol. 159, 2009, JENAM 2008 Symposium № 4: Asteroseismology and Stellar Evolution
S. Schuh & G. Handler

"Hybrid" pulsators - fact or fiction?

G. Handler

Institut für Astronomie, Universität Wien, Türkenschanzstraße 17, 1180 Vienna, Austria

Abstract

We carried out a multi-colour time-series photometric study of six stars that were claimed as "hybrid" p- and g-mode pulsators in the literature. The β Cep/SPB star γ Peg was confirmed and revealed excellent asteroseismic potential. HD 8801 was confirmed as a "hybrid" δ Sct/γ Dor star; additional pulsation frequencies were detected. 53 Psc likely is an SPB star and the O-type star HD 13745 showed small-amplitude slow variability. No light variations were detected for HD 19374 and, surprisingly, ι Her.

Individual Objects: γ Peg, 53 Psc, HD 8801, HD 13745, HD 19374, ι Her

Introduction

Stars can self excite observable pulsations if an excitation mechanism operates in a resonant cavity in their interior or on their surface. Since this is only fulfilled under certain physical conditions, pulsating stars are located in different instability domains in the HR diagram. These instability strips are not necessarily distinct. Stars having two different sets of pulsational mode spectra excited simultaneously may therefore exist within overlapping instability strips. This is good news for asteroseismology, as both types of oscillation can be exploited to obtain a more complete picture of the stellar interior.

Following this idea, Handler et al. (2002) discovered a star that showed both γ Doradus type g modes and δ Scuti type p modes. However, at least some of the g modes may have been excited through tidal effects from a close companion in an eccentric orbit. Nevertheless, following this discovery several such "hybrid" pulsators were reported in the literature. Henry & Fekel (2005) discovered both γ Doradus and δ Scuti type pulsations in a single Am star, and the MOST satellite found two additional examples (King et al. 2006, Rowe et al. 2006), both of which are again Am stars.

Among the B type stars, "hybrid" SPB/β pulsations have been reported for several objects (e.g., see Jerzykiewicz et al. 2005, Handler et al. 2006, Chapellier et al. 2006, De Cat et al. 2007). In addition, two subdwarf B stars were also discovered to show "hybrid" oscillations (Oreiro et al. 2005, Baran et al. 2005, Schuh et al. 2006). The main physical difference between the B type and A/F type "hybrid" pulsators is that in the first group the same driving mechanism excites both types of oscillation, whereas in the δ Sct/γ Dor stars two different driving mechanisms are at work.

Besides all the exciting possibilities that "hybrid" pulsators offer for asteroseismology, the major observational problem that needs to be solved before arriving at a unique seismic model still remains the same - or is even more severe: a sufficiently large number of pulsation modes must be detected and identified - in *both* frequency domains. Consequently, the best targets for asteroseismic studies need to be identified before embarking on large-scale projects. To this end, we have selected six stars that have been claimed as "hybrid" pulsators in the literature for such an exploratory study, comprising one main sequence A/F star and five O/B stars.

Observations and results

We carried out time-series (u)vy photometry of these six stars of the 0.75-m Automatic Photoelectric Telescope (APT) T6 at Fairborn Observatory in Arizona, between October 2007 and June 2008. An overview of the observations and the results is given in Table 1.

Table 1: Results of our photometric survey. ΔT is the time span of the data set, T_{tot} is the total number of hours observed, N_{tot} is the number of nights observed, and N_{obs} is the number of data points obtained.

Star	ΔT	T_{tot}	N_{tot}	N_{obs}	Filters	Classification
γ Peg	69	284	48	884	uvy	"hybrid" SPB/β Cep star
53 Psc	69	281	48	861	uvy	SPB star
HD 8801	69	272	48	737	vy	"hybrid" γ Dor/δ Sct star
HD 13745	69	243	47	616	uvy	slowly variable
53 Ari	69	283	48	927	uvy	not found to vary
ι Her	78	118	45	498	uvy	not found to vary

HD 8801 was confirmed as a "hybrid" γ Dor/δ Sct star, as were all frequencies found in the discovery data (Henry & Fekel 2005), allowing for some aliasing ambiguities. In particular, we confirm the presence of several frequencies intermediate between the γ Dor and δ Sct domains. Their origin remains to be understood; an explanation in terms of binarity is unlikely given that HD 8801 seems to be a single star. Additional pulsation frequencies were detected in the two lower-frequency domains in our data.

Our measurements of 53 Psc are consistent with two close frequencies around 0.87 cycles/day. Their uvy amplitudes are consistent with an interpretation in terms of low-order g-mode pulsation. No β Cep-type pulsations were detected within a limit of 0.8 mmag. We note that the presence of β Cep pulsation of 53 Psc has already been disputed in the literature (Le Contel et al. 2001, De Cat et al. 2007), but we observed it anyway due to its proximity to γ Peg in the sky.

The O-type star HD 13745 was claimed to be a "hybrid" β Cep/SPB star by De Cat et al. (2007), which would be particularly interesting given its spectral classification. However, our measurements only showed a complex, low-amplitude variation with a time scale of $\sim$ 3.2 d, and no evidence for β Cep pulsation within a limit of about 1 mmag.

The stars 53 Ari and ι Her showed no discernible variability in our measurements, within a (generous) 1.5 mmag limit. This is particularly surprising for ι Her because this star has repeatedly, and convincingly, been reported as variable in the literature (Chapellier et al. 2000 and references therein).

Our most encouraging result was the clear confirmation of the "hybrid" nature of γ Peg; four g modes and two (previously known) p modes were detected. We took this as a motivation for an extended observational effort on the star, involving high-precision photometry with the MOST satellite aided by high-resolution multisite spectroscopy and ground-based multicolour photometry to facilitate pulsational mode identification. Preliminary results indicate the presence of additional g modes and of a virtually complete set of $l = 0 - 2$ modes in the domain of excited β Cephei-type modes.

Acknowledgments.   This work is supported by the Austrian Fonds zur Förderung der wissenschaftlichen Forschung under grant P20526-N16.

References

Baran, A., Pigulski, A., Koziel, D., et al. 2005, MNRAS, 360, 737

Chapellier, E., Mathias, P., Le Contel, J.-M., et al. 2000, A&A, 362, 189

Chapellier, E., Le Contel, D., Le Contel, J.-M., et al. 2006, A&A, 448, 697

De Cat, P., Briquet, M., Aerts, C., et al. 2007, A&A, 463, 243

Handler, G., Balona, L. A., Shobbrook, R. R., et al. 2002, MNRAS, 333, 262

Handler, G., Jerzykiewicz, M., Rodríguez, E., et al. 2006, MNRAS, 365, 327

Henry, G. W., & Fekel, F. C. 2005, AJ, 129, 2026

Jerzykiewicz, M., Handler, G., Shobbrook, R. R., et al. 2005, MNRAS, 360, 619

King, H., Matthews, J. M., Rowe, J. F., et al. 2006, CoAst, 148, 28

Le Contel, J.-M., Mathias, P., Chapellier, E., & Valtier, J.-C. 2001, A&A, 380, 277

Oreiro, R., Pérez Hernández, F., Ulla, A., et al. 2005, A&A, 438, 257

Rowe J. F., Matthews J. M., Cameron C., et al. 2006, CoAst, 148, 34

Schuh, S., Huber, J., Dreizler, S., et al. 2006, A&A, 445, L31

Comm. in Asteroseismology,
Vol. 159, 2009, JENAM 2008 Symposium № 4: Asteroseismology and Stellar Evolution
S. Schuh & G. Handler

TW Dra: NRP mode identification with FAMIAS

H. Lehmann,[1] A. Tkachenko,[1] and D. E. Mkrtichian [2,3]

[1] Thüringer Landessternwarte Tautenburg, Germany
[2] ARCSEC, Sejong University, Seoul, Korea
[3] Astronomical Observatory, Odessa National University, Ukraine

Abstract

TW Dra is an Algol type system where the primary shows δ Sct oscillations. Time series of high-resolution spectra taken at the Thüringer Landessternwarte Tautenburg (TLS) show a complex pattern of moving bumps across the line profiles, indicating a rich spectrum of high-degree non-radial pulsation modes. After the reductions to remove the signatures of the second and third components we analyze the line profiles using the FAMIAS program to derive the oscillation frequencies and to identify the pulsation modes. We describe the way in which mode identification can be done by FAMIAS and present first results.

Individual Objects: TW Dra

Introduction

The Algol type system TW Dra (A5V+K0III) was detected to be a member of the new class of oEA stars (i.e. Algol type systems where the mass accreting primary shows δ Scuti oscillations; Mkrtichian et al. 2002, 2004) by Kusakin et al. (2001). Photometrically, it is one of the best investigated oEA stars. Here we present a first spectroscopic investigation of non-radial pulsation (NRP) modes using the program package FAMIAS (Zima 2008). Time series of high-resolution spectra were taken in 13 consecutive nights with the Coudé-echelle spectrograph at the 2m telescope at TLS and in two nights with the BOES spectrograph at the 1.8m telescope at Bohyunsan Optical Astronomy Observatory (BOAO).

The analysis is complicated by the fact that TW Dra is a member of a visual binary and that the obtained spectra partly contain light contributions from the third component. In preparation of the investigation of NRP modes we derived first a precise orbital solution using KOREL (Hadrava 2004). From this we obtained the masses M_1=2.13 $M_\odot$, M_2=0.89 $M_\odot$, and the separation a=12.11 $R_\odot$ of the components of the close system. KOREL delivered also the disentangled spectra of all three components from which we derived important parameters of the primary like $v \sin i$=47 km s^{-1}, microturbulence of 2.9 km s^{-1} and a slight metal overabundance. Values of T_{eff}=8150 K and $\log g$=3.88 were taken from the photometric solution. To gain a higher S/N we applied least squares deconvolution (LSD, Donati et al. 1997) to the spectra and removed the signatures of the second and third components in fitting the disentangled spectra shifted by the orbital radial velocity to the composite LSD line profiles. All these procedures and results are described by Lehmann et al. (2008).

Frequency analysis of oscillation modes and mode identification

We did a multiple frequency search based on successive pre-whitening of the data as well in the line profile moments as in the profiles itself by using the pixel-by-pixel method (calculation of the Fourier spectrum of the temporal variation of each pixel-value across the line profile). Here we restricted the used dispersion range as close as possible to the inner line core where the high-frequency contributions dominate while the variations in the outer wings are mainly determined by low frequencies arising from orbital motion and imperfect removal of the second and third components' contributions. By using the Fourier method, three oscillations in the high-frequency domain were found: $F_1=22.90\,\mathrm{cd}^{-1}$, $F_2=14.05\,\mathrm{cd}^{-1}$, and $F_3=24.72\,\mathrm{cd}^{-1}$, in difference to the values of $17.99\,\mathrm{cd}^{-1}$ or $18.95\,\mathrm{cd}^{-1}$ that were detected by Kusakin et al. (2001) or Kim et al. (2003) photometrically. Also a frequency search in the first three LSD profile moments did not reveal the photometric values, but in the second order moment we found a narrow doublet centred at this place. Instead we found two frequencies of $f_1=20.29\,\mathrm{cd}^{-1}$ and $f_2=25.26\,\mathrm{cd}^{-1}$ or their $1\,\mathrm{cd}^{-1}$ or $2\,\mathrm{cd}^{-1}$ aliases in all of the three moments. The $1\,\mathrm{cd}^{-1}$ alias of f_1 is also the strongest mode detected in the photometry obtained by E. Rodriguez at Sierra Nevada Observatory in three nights in March 2008 where the f_2 mode can be seen as well.

Based on the multiple frequency model obtained from the Fourier analysis of the line profiles, we tried to identify the modes by using the Fourier Parameter Fit (FPF) method of FAMIAS. The model included the three oscillation frequencies F_1 to F_3 as well as some low-frequency contributions. Basic stellar parameters were fixed to the above mentioned values. In a first step we derived the equivalent width and RV zero point of the mean profile. After fixing the pulsation phases found from a free parameter search, we used the grid-mode search to optimize l and m quantum numbers together with the remaining free parameters $v \sin i$, intrinsic width and pulsation velocity. Unfortunately, we did not find an unique solution and present here the solutions having lowest chi-square: $(l, m) = (7,5)$, $(7,7)$, $(8,8)$ for F_1; $(7,5)$, $(8,6)$, $(8,8)$ for F_2; $(10,6)$, $(11,7)$, $(11,9)$, $(11,11)$ for F_3.

Conclusions

Using the pixel-by-pixel method of FAMIAS, three oscillation modes in the high-frequency domain have been detected which are all different from those found in the photometry before. We found no unique solution from mode identification and can present here only different possible combinations of l and m. All these combinations include only high-degree l modes. We have also detected two pulsation modes in the first three moments which are in agreement with those found in the photometry. We assume that from the line profile moments and from photometry we will find only low degree modes whereas the high-degree l modes can be found only with the pixel-by-pixel method.

In the near future we want to improve the removal of the secondary and third components with the aim to use broader line regions in the analysis, to analyze the suspected low-degree modes, and to model the eclipse phases with SHELLSPEC (Budaj et al. 2005) to analyze line profiles variations by eclipse mapping (Gamarova et al. 2003).

Acknowledgments. Mode identification was obtained with the software package FAMIAS developed within the FP6 European Coordination Action HELAS (http://www.helas-eu.org/).

References

Budaj, J., Richards, M. T., & Miller, B. 2005, ApJ, 623, 411

Donati, J.-F., Semel, M., Carter, B. D., et al. 1997, MNRAS, 291, 658

Gamarova, A. Yu., Mkrtichian, D. E., Rodriguez, E., et al. 2003, ASP Conf. Ser., 292, 369

Hadrava, P. 2004, Publ. Astron. Inst. ASCR, 92, 15

Kim, S.-L., Lee, J. W., Kwon, S.-G., et al. 2003, A&A, 405,231

Kusakin, A. V., Mkrtichian, D. E., & Gamarova, A. Yu. 2001, IBVS, 5106

Lehmann, H., Tkachenko, A., Tsymbal, V., & Mkrtichian, D. E. 2008, CoAst, 157, 332

Mkrtichian, D. E., Kusakin, A. V., Rodriguez, E., et al. 2004, A&A, 419, 1015

Mkrtichian, D. E., Kusakin, A. V., Gamarova, A. Yu., & Nazarenko, V. 2002, ASP Conf. Ser., 259, 96

Zima, W. 2008, CoAst, 155, 17

Excursion on the Danube river.

Comm. in Asteroseismology,
Vol. 159, 2009, JENAM 2008 Symposium № 4: Asteroseismology and Stellar Evolution
S. Schuh & G. Handler

Preliminary results of V440 Per and α UMi observations with the Poznan Spectroscopic Telescope

M. Fagas, R. Baranowski, P. Bartczak, W. Borczyk, W. Dimitrow, K. Kaminski,
T. Kwiatkowski, R. Ratajczak, A. Rozek, and A. Schwarzenberg-Czerny

Obserwatorium Astronomiczne, Uniwersytet im. Adama Mickiewicza, ul. Słoneczna 36, 60-286 Poznań,
Poland

Abstract

Presented herein are preliminary radial velocity results for two classical Cepheids: V440 Per
and α UMi (Polaris). Both stars have been observed with the Poznan Spectroscopic Telescope (PST), operational since Aug 2007 at the Borowiec Station of Poznań Astronomical
Observatory in Poland. Data obtained for V440 Per suggest the presence of a low-amplitude
secondary mode of pulsations. Results of α UMi observations confirm further pulsation amplitude growth, as observed during the last decades.

Individual Objects: V440 Per, Polaris

Introduction

Both V440 Per and α UMi spectroscopy has been obtained with the old PST configuration.
The telescope was equipped with two 0.4 m parabolic mirrors (only one of them was operational during the period of observations). PST is currently being adjusted to simultaneously
use both mirrors, which have been replaced with 0.5 m ones in August 2008.

PST is equipped with a mid-resolution ($R = 35000$) fibre-fed echelle spectrograph, based
on the MUSICOS construction design (Baudrand & Böhm 1992). It covers the spectral range
between 4500 and 9200 Å. Image acquisition was performed with a high-quality low-noise
2k × 2k ANDOR CCD, at a temperature of $-90°$ C.

Data reduction and radial velocity measurements were done using the IRAF package
(http://iraf.noao.edu/), as well as with combined Python and cl scripts developed by our
team members. The cross-correlation method was used to obtain radial velocities. Further
periodicity analysis was conducted using the Period04 (Lenz & Breger 2005) and TATRY
(Schwarzenberg-Czerny 1996) software.

V440 Per

V440 Per is a bright (6.3 mag) star of F7 Ib spectral type. It is an overtone pulsator, with an
oscillation period of 7.570 d (Luck et al. 2008). 158 spectra for this star have been obtained
with PST between August 2007 and July 2008. Typical exposure times were $600 - 900$ s,
with a signal-to-noise ratio (S/N) varying from $\sim 20 - 40$, depending on weather conditions.
Radial velocity data are presented in Fig. 1.

Preliminary radial velocity measurements revealed a low-amplitude (~ 90 m/s) second
harmonic of V440 Per pulsations. If confirmed, that would make V440 Per the overtone

V440 Per
PST, 14 Aug 2007 - 03 Jul 2008, P = 7.57 d, HJD0 = 2454325.202

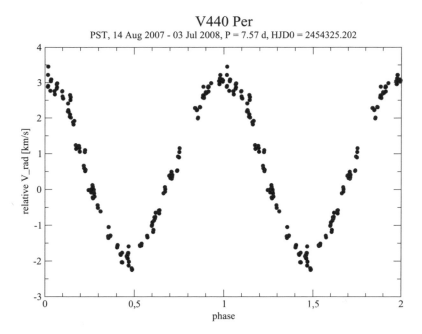

Figure 1: Radial velocity curve of V440 Per obtained with PST.

Cepheid with the longest pulsation period known. Data analysis is being carried out in cooperation with Pawel Moskalik.

α UMi (Polaris)

Spectroscopic and photometric observations of this bright classical Cepheid over the last century revealed changes in both pulsation period and amplitude. While the period was increasing, the amplitude (both photometric and spectroscopic) seemed to decrease, leading to the conclusion that Polaris' pulsations might eventually cease. This trend, however, diverted during the last decades (Lee et al. 2008).

PST's design allows easy observations of Polaris. 330 spectra have been taken over seven months (Dec 2007 - Jul 2008) with a typical exposure time of 150 s. The S/N in the obtained spectra was ∼ 30 − 70, depending on weather conditions.

190 α UMi spectra (Dec 2007 - Apr 2008) have been reduced so far. The influence of Polaris' orbital motion on the star's pulsation period has not been taken into account, as the analyzed data set only covers 4 months (∼ 1% of the orbital period). Preliminary analysis of the radial velocity data revealed a pulsation period slightly longer than expected $(3.978 \pm 0.002\,d)$. The 2K amplitude found was 2.52 ± 0.03 km/s, which is consistent with recent work by Lee et al. (2008) and Spreckley & Stevens (2008).

The star is scheduled for further observations with PST in the upcoming years.

Acknowledgments. The authors acknowledge support from the Polish MNiI/MNiSzW Grant 1 P03D 025 29.

References

Baudrand, J., & Bohm, T. 1992, A&A, 259, 711

Lee, B. C., Mkrtichian, D. E., Han, I., et al. 2008, AJ, 135, 2240

Lenz, P., & Breger, M. 2005, CoAst, 146, 53

Luck, R. E., Andrievsky, S. M., Fokin, A., & Kovtyukh, V. V. 2008, AJ, 136, 98

Moskalik, P., private communication

Schwarzenberg-Czerny, A. 1996, ApJ Letters, 460, L107

Spreckley, S. A., & Stevens, I. R. 2008, MNRAS, 388, 1239

Excursion on the Danube river.

Comm. in Asteroseismology,
Vol. 159, 2009, JENAM 2008 Symposium № 4: Asteroseismology and Stellar Evolution
S. Schuh & G. Handler

2D modeling of a Cepheid, moving grid approach

E. Mundprecht

Fakultät für Mathematik, Universität Wien, Nordbergstrasse 15, 1090 Vienna, Austria

Abstract

We present the first 2D simulation of a Cepheid of an effective temperature of 5125K.

The hydrodynamic equations in polar coordinates and their implementation in the ANTARES code

The equations governing the hydrodynamics are the continuity equation, the momentum density equation and the "total" energy density equation. Due to the size of the computational domain they are evolved in polar coordinates.

Continuity:
$$\partial_t \rho = -\nabla \cdot \left(\vec{l} - \rho \vec{u} \right)$$

Energy:
$$\partial_t E = -\left(\nabla \cdot \left(\frac{\vec{l}}{\rho} (E + p - \sigma') - E\vec{u} \right) + \vec{l} \cdot \vec{g} + Q \right)$$

Momentum in radial direction: $\partial_t l_r =$

$$-\left(\partial_r \left(\frac{l_r^2}{\rho} - u l_r + p - \sigma'_{11} \right) + \frac{1}{r}\partial_\theta \left(\frac{l_r l_\theta}{\rho} - \sigma'_{21} \right) - \frac{2 l_r^2}{\rho r} + \frac{2 l_r u}{\rho} + \frac{l_\theta^2}{\rho r} + \frac{l_r l_\theta}{r\rho} \cot\theta + \sigma_{corr_r} + \rho g \right)$$

Momentum in angular direction: $\partial_t l_\theta =$

$$-\left(\partial_r \left(\frac{l_r l_\theta}{\rho} - u l_\theta - \sigma'_{12} \right) + \frac{1}{r}\partial_\theta \left(\frac{l_\theta^2}{\rho} + p - \sigma'_{22} \right) - 3 \frac{l_r l_\theta}{r\rho} + \frac{2 l_\theta u}{\rho} + \frac{l_\theta^2}{r\rho} \cot\theta + \sigma_{corr_\theta} \right)$$

with the momentum vector $\vec{l} = (l_r, l_\theta)$. In every vector, the first component denotes the radial direction, and the second the angular direction. The other variables are: the grid velocity $\vec{u} = (u, 0)$, the gas pressure p, and $\vec{g} = (g, 0)$ gravity. The radiative heating rate Q is determined by the short-characteristic method (in the outermost region) and diffusion approximation, respectively. σ' is the viscous tensor in polar form.

All fluxes are computed at the cell centre and interpolated to the cell boundaries. For the hyperbolic part an ENO scheme with Marquina flux splitting (Fedkiw et al. 1996) is used. During the updating process the difference in cell volume due to the moving grid is included in the divergences and derivatives of the above equations. At the end of each time step the grid is advanced. The grid velocity u^{top} at the top is determined as the horizontal average of the fluid velocity. The outermost grid point moves with u^{top}, the innermost grid point remains fixed, at the intermediate points the grid velocity and the grid points are computed via a dilatation factor. For details on the ANTARES code see Muthsam et al. (2007).

Up to now two models have been investigated, one with $T_{eff} = 5125\,\text{K}$ and one with $T_{eff} = 5500\,\text{K}$. Here, I present the cooler Cepheid. Some results including a bolometric light curve of the other star are presented by H. Muthsam in the proceedings of symposium 7 of this JENAM conference.

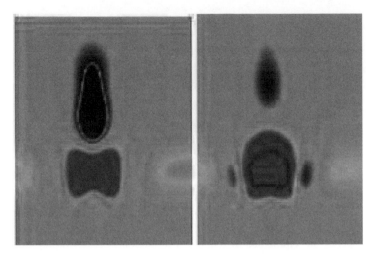

Figure 1: Convective fluxes F_c: upwards red (black contours), downwards blue (white contours). Only one of nine convective cells at the very top of the computational domain is depicted. Left: contracted, right: expanded.

Physical properties of the star and the domain

Main Properties: $M = 5\,M_\odot$, $L = 912.797\,L_\odot$, $\log g = 1.96$, $R = 6.75\,\text{Gm}$, $P = 4.16\,\text{d}$, $(X, Y, Z) = (0.737, 0.243, 0.02)$.

The computational domain reaches from $T = 4600\,\text{K}$ to $T = 320000\,\text{K}$ and covers the top 42% of the star, with an aperture angle of 30° and 450 horizontal grid points. An evenly stretched grid of 380 vertical grid points is a applied on the radial component, the cell height varies from $1.98\,\text{Mm}$ at the top to $125.3\,\text{Mm}$ at the bottom of the domain.

The simulation was started in 1D, after approx. 10 days regular pulsations were established. After 56 days a randomly perturbed angular momentum was added. The 2D simulation is now 34 days old. While the perturbations in the convective flux $F_c = I_r'\,\rho'^{-1}\,(\rho h)'$ still grow over time, correlations between the convective pattern and the phase of the pulsation are clearly visible. During the expansion the upwards flow increases and reaches a maximum shortly after the maximum radius, the downwards flow decreases. During the contraction the downwards flow increases, the upwards flow decreases. At the maximum expansion the temperature contrast between up- and down-flowing material is $\sim 890\,\text{K}$ and thus smaller as at total contraction ($\sim 1400\,\text{K}$). Also, the topology varies.

Acknowledgments. I am indebted to the Austrian Science Foundation for grants P18224 and P20762. G. Houdek, Univ. of Cambridge, kindly provided the hydrostatic starting model of the Cepheids and linear stability data.

References

Fedkiw, R. P., Merriman, B., Donat, R., & Osher, S. 1996, Proceedings of the symposium on progress in numerical solutions of partial differential equations, Arcachon, France

Muthsam, H. J., Löw-Baselli, B., Obertscheider, C., et al. 2007, MNRAS, 380, 1335

Comm. in Asteroseismology,
Vol. 159, 2009, JENAM 2008 Symposium № 4: Asteroseismology and Stellar Evolution
S. Schuh & G. Handler

Blazhko variables – recent results

J. Jurcsik

Konkoly Observatory of the Hungarian Academy of Sciences, P.O. Box 67, H-1525 Budapest, Hungary

Abstract

The light curve modulation of a fraction of the RR Lyrae stars is still one of the unexplained phenomena of stellar pulsation. The suggested models cannot explain even the most prominent observed properties of the light variation.

During the past five years 90% of the observing time on the 60 cm automatic telescope of Konkoly Observatory was dedicated to observing RR Lyrae stars, especially those showing Blazhko modulation. In the course of the Konkoly Blazhko Survey we have obtained the densest and most extended standard multicolour photometric observations of about a dozen Blazhko variables. Detailed analyses of our observations of RR Gem, SS Cnc and MW Lyrae have already been published (Jurcsik et al. 2006a, 2006b, 2008), while similar studies of many other Blazhko variables are in progress. These data allow us to detect previously unknown properties of the modulation which help to gain more complete insight into the phenomenon.

Some of the most important new achievements obtained in the course of our studies are the following:

- Contrary to previous results, which described the light variation with triplet frequencies at the main pulsation component and its harmonics, we detected quintuplet and even more complex modulation frequency patterns in the Fourier spectra of Blazhko variables (Hurta et al. 2008, Jurcsik et al. 2008).

- Contrary to previous assumptions that the amplitude of the modulation is relatively large in all Blazhko variables we detected small amplitude modulations as well (Jurcsik et al. 2006a, 2006b).

- Contrary to previous results we detected modulations with periods as short as some days (Jurcsik et al. 2006a, 2006b). This result warns that those automatic selections of Blazhko variables which search for modulation frequencies e.g., only in the 0.1 c/d vicinity of the pulsation frequencies (e.g., Collinge et al. 2006) are biased by artificial omission of short period modulations.

- Our extended multicolour photometric data of Blazhko variables which fully cover the pulsation light curves in different phases of the modulation make it possible to determine the changes in the global mean properties of the stars connected with the modulation cycle (Jurcsik et al. 2009) using an inverse photometric Baade-Wesselink method developed by Sódor et al. (2009).

These results may have crucial importance in finding the correct physical explanation of the Blazhko effect more than hundred years after its discovery.

Individual Objects: RR Gem, SS Cnc, MW Lyrae

Acknowledgments.   The financial support of OTKA grants T-068626 and T-048961 is acknowledged.

References

Collinge, M., Sumi, T., & Fabrycky, D. 2006, ApJ, 651, 197

Hurta, Zs., Jurcsik, J., Szeidl, B., & Sódor, Á. 2008, AJ, 135, 957

Jurcsik, J., Sódor, Á., Váradi, M., et al. 2006a, A&A, 430, 104

Jurcsik, J., Szeidl, B., Sódor, Á., et al. 2006b, AJ, 132, 61

Jurcsik, J., Sódor, Á., Hurta, Zs., et al. 2008, MNRAS, 391, 164

Jurcsik, J., Sódor, Á., Szeidl, B., et al. 2009, MNRAS, 393, 1553

Sódor, Á., Jurcsik, J., & Szeidl, B. 2009, MNRAS, 394, 261

Comm. in Asteroseismology,
Vol. 159, 2009, JENAM 2008 Symposium № 4: Asteroseismology and Stellar Evolution
S. Schuh & G. Handler

CZ Lacertae – a Blazhko RR Lyrae star with multiperiodic modulation

Á. Sódor

Konkoly Observatory of the Hungarian Academy of Sciences, P.O. Box 67, H-1525 Budapest, Hungary

Abstract

CZ Lacertae, a fundamental mode RR Lyrae star, was observed during the Konkoly Blazhko Survey at Konkoly Observatory. The observations covered two consecutive seasons and were obtained with a Wright CCD camera attached to the 60 cm automatic telescope at Svábhegy, Budapest. The log of the multicolour ($BVR_C I_C$) observations are summarized in Table 1.

After several nights of observation it was obvious that CZ Lac was a Blazhko star and its modulation was rather complex. Therefore we collected an extended amount of data on this object. Our Konkoly Blazhko Survey provides the most extended multicolour Blazhko observations ever obtained; the data of CZ Lac and MW Lyr (Jurcsik et al. 2008, 2009) are the most numerous. The V light curve of the observations from the two seasons folded with the pulsation period is shown in Fig. 1.

Our data show that this variable is modulated with two different periods. The two modulations have similar amplitudes i.e., none of them is dominant. The multiperiodic behaviour of the modulation is hard to be explained by those theoretical models of the phenomenon that bind the modulation period to the rotation of the star e.g., the oblique magnetic rotator model.

The multiperiodic modulation of RR Lyrae stars is not a completely new phenomenon, as it was suspected earlier e.g., in XZ Cyg (LaCluyzé et al. 2004) or at UZ UMa (Sódor et al. 2006). Also the Blazhko RR Lyrae stars of the MACHO and OGLE surveys (Alcock et al. 2000, Moskalik & Poretti 2003) that have unequally spaced triplet structures in their Fourier spectra are possibly multiperiodically modulated variables. Nonetheless, CZ Lac is the first Blazhko star with multiperiodic modulation where both modulation periods are unambiguously identified. The length of the data set allows to resolve the modulations in both seasons independently. Both modulation frequencies show significant changes between the two seasons, which suggests that multiperiodic modulations may have a greater instability than monoperiodic ones. In contrast with CZ Lac, MW Lyr with a single Blazhko period shows very stable modulation throughout two seasons of observations (20 Blazhko cycles).

There are complex multiplet structures around the pulsation harmonics (kf_0) in the Fourier spectra. Not only modulation sidelobes ($kf_0 \pm f_{m_1}$ and $kf_0 \pm f_{m_2}$) with the two modulation frequencies (f_{m_1} and f_{m_2}) appear but also sidelobes with their linear combinations ($kf_0 + (f_{m_1} + f_{m_2})$ and $kf_0 \pm (f_{m_1} - f_{m_2})$), and even with a subharmonic of the larger frequency ($kf_0 \pm 0.5 f_{m_2}$) can be observed around several pulsation harmonics.

A detailed study on the analysis of the Blazhko behaviour of CZ Lac is in preparation and is planned to be published elsewhere.

Individual Objects: CZ Lac

Table 1: Log of our CCD multicolour $BVR_C I_C$ observations of CZ Lac.

Season	Begin [JD]	Length [d]	No. of data points per band
2004 – 2005	2 453 266	146	≈ 3650
2005 – 2006	2 453 648	84	3011 – 4460

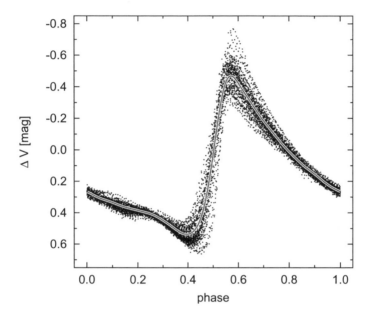

Figure 1: V light curve of CZ Lac, folded with the pulsation period, with the fitted mean curve.

Acknowledgments.   The financial support of OTKA grants T-068626 and T-048961 is acknowledged.

References

Alcock, C., Allsman, R., Alves, D. R., et al. 2000, ApJ, 542, 257

Jurcsik, J., Sódor, Á., Hurta, Zs., et al. 2008, MNRAS, 391, 164

Jurcsik, J., Sódor, Á., Szeidl, B., et al. 2009, MNRAS, 393, 1553

LaCluyzé, A., Smith, H. A., Gill, E.-M., et al. 2004, AJ, 127, 1653

Moskalik, P., & Poretti, E. 2003, A&A, 398, 213

Sódor, Á., Vida, K., & Jurcsik, J. 2006, IBVS, 5705

Comm. in Asteroseismology,
Vol. 159, 2009, JENAM 2008 Symposium № 4: Asteroseismology and Stellar Evolution
S. Schuh & G. Handler

The true Blazhko behaviour of DM Cyg

Zs. Hurta [1,2]

[1] Eötvös Loránd University, Department of Astronomy, Budapest, Hungary
[2] Konkoly Observatory, Budapest, Hungary

Abstract

We present preliminary results of our work on DM Cyg, an RRab star with steadily increasing pulsation period. The Blazhko modulation of the light curve of DM Cyg has not been undoubtedly confirmed yet. A reanalysis of the original data (Sódor & Jurcsik 2005) could not confirm the 26 d periodicity found by Lysova & Firmanyuk (1980) in the timings of maximum brightness data of visual observations. Neither the scarce photoelectric observations (Fitch 1966, Sturch 1966, Hipparcos 1997) nor the CCD data of the NSVS (Woźniak 2004) survey suggested a notable light curve modulation.

In order to get a definite answer whether the light curve of DM Cyg is stable or it shows any kind of modulation it was observed in the course of the Konkoly Blazhko Survey in the 2007 and 2008 seasons. Using the automated 60 cm telescope of the Konkoly Observatory, Svábhegy, Budapest, equipped with a Wright 750 x 1100 CCD camera and BVI_C filters we obtained more than 3000 data points on about 80 nights in each band. Archive photoelectric and photographic observations obtained with the 60 cm telescope and a 16 cm astrograph of the Konkoly Observatory in 1978 and between 1934 and 1958 were also analyzed. The photoelectric and photographic photometry provided 75 B,V and 1031 pg data points from 4 and 40 nights, respectively.

The CCD observations revealed that the light curve of DM Cyg is in fact modulated, but with very small amplitude. The maximum brightness variation hardly exceeds 0.05 mag in the V band, while no definite phase modulation of the light curve and/or maximum timings is evident. The amplitudes of the modulation frequencies that form equidistant triplets around the pulsation frequency and its harmonics are below 15 mmag. There is some indication of light curve modulation in the Konkoly photographic data as well.

Our data confirm that DM Cyg shows Blazhko modulation but with significantly different period and character (amplitude/phase modulation) than it was found by Lysova & Firmanyuk (2000). A detailed analysis of our observations of DM Cyg with its true Blazhko period will be submitted to MNRAS in early 2009.

Individual Objects: DM Cyg

Acknowledgments. The financial support of OTKA grants K-68626 and T-048961 is acknowledged.

References

Fitch, W. S., Wisniewski, W. Z., & Johnson, H. L. 1966, Comm. Lunar and Planetary lab., 5, 71

Lysova, I , & Firmanyuk, V. 1980, Ast. Cir., 1122, 3

Sódor Á., & Jurcsik, J. 2005, IBVS, 5641

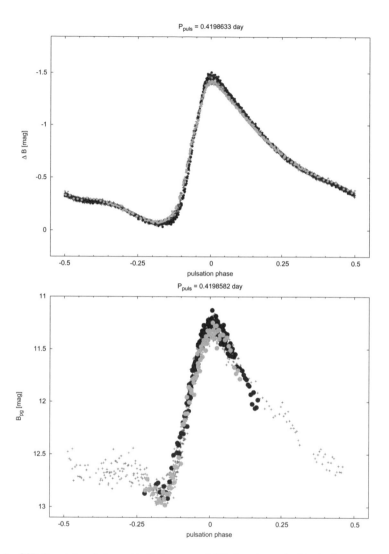

Figure 1: CCD B and the photographic light curves of DM Cyg observed with the instruments of the Konkoly Observatory. Data belonging to the largest and smallest amplitude phases of the modulation are plotted with black and grey dots, respectively. The CCD data show without doubt that the light curve of DM Cyg is not stable; the full amplitude of the pulsation in B band varies by about 0.1 mag. There is some indication that the modulation of the light curve is also present in the photographic data, but taking into account the higher noise of the photographic data and also the small amplitude of the modulation, the photographic data alone are not suitable to determine the Blazhko behaviour of DM Cyg.

Sturch, C. 1966, ApJ, 143, 774

Woźniak, P. R., Vestrand, W. T., Akerlof, W. C., et al. 2004, AJ, 127, 2436

Comm. in Asteroseismology,
Vol. 159, 2009, JENAM 2008 Symposium № 4: Asteroseismology and Stellar Evolution
S. Schuh & G. Handler

Asteroseismology of pre-main sequence stars

K. Zwintz,[1] D. B. Guenther,[2] and T. Kallinger[1]

[1] Institut für Astronomie, Universität Wien, Türkenschanzstrasse 17, 1180 Vienna, Austria
[2] Department of Astronomy and Physics, St. Mary's University, Halifax, NS B3H 3C3, Canada

Abstract

Intermediate-mass pre-main sequence (PMS) stars cross the instability region on their way to the main sequence and can become pulsationally unstable. The δ Scuti-like PMS pulsators populate the same area in the Hertzsprung-Russell (HR) diagram as the classical δ Scuti stars. But the interiors of PMS stars differ from those of their more evolved (post-)main sequence counterparts resulting in different pulsation frequency spectra. Asteroseismology of PMS p-mode pulsators has only recently become possible. The latest investigation focused on the analysis of MOST space photometry of the Herbig Ae star HD 142666 finding 12 pulsational frequencies lying on top of larger irregular variability caused by a circumstellar disk.

Individual Objects: HD 142666

Introduction

Young stars mainly gain their energy from gravitational contraction before they arrive on the zero-age main sequence (ZAMS). Their interiors show different density profiles compared to their (post-)main sequence counterparts (Marconi & Palla 1998) and lack regions of already processed nuclear material. Such pre-main sequence (PMS) objects cross the instability region in the HR diagram while evolving to the ZAMS and can become pulsationally unstable. As the stellar density profiles in PMS stars are different than in (post-)main sequence stars of same effective temperature, luminosity and mass, the pulsational frequency spacings also differ for the two evolutionary stages (Suran et al. 2001). Hence, with a dense enough pulsational frequency spectrum, asteroseismology can be used to investigate the evolutionary stage of a field star.

The PMS Instability Strip

For the 36 known PMS pulsators (i.e., 18 Herbig Ae field stars and 18 members of young open clusters), a comparison to the classical δ Scuti stars was conducted (Zwintz 2008). While a wealth of astrophysical parameters are available for the classical δ Scuti objects, the only common parameters of the 36 pulsating PMS stars are Johnson V and $(B - V)$ or their corresponding dereddened values of M_V and $(B - V)_0$. The comparison of these pulsators in the observed HR diagram (in the M_V vs. $(B - V)_0$ plane) showed that δ Scuti stars of both evolutionary stages occupy the same instability region (Zwintz 2008).

HD 142666

Only one PMS pulsator is located redwards of the red boundary of the classical δ Scuti instability strip: HD 142666. This Herbig Ae star (spectral type A8 Ve and $V = 8.81$ mag) is known to have a dense circumstellar dust disk seen nearly edge-on (e.g., Meeus et al. 1998) that causes irregular brightness variations of up to a magnitude peak-to-peak which are accompanied by colour changes. When the dust clouds are in the observer's line-of-sight, they obscure the star and cause the reddening. As the star is never seen free from its circumstellar dust, its $(B - V)_0$ can only be determined with large relative errors which explains HD 142666's "extreme" location in the M_V vs. $(B - V)_0$ plane.

Kurtz & Müller (2001) discovered a single pulsation frequency of 21.43 d^{-1} (i.e., period of 1.12 hours) for HD 142666. In 2006 and 2007 the MOST (Microvariability and Oscillations of Stars) satellite (Walker et al. 2003) observed HD 142666 for 11.5 days and 39 days, respectively. 12 pulsational frequencies were identified in the data sets of both years and were submitted to a detailed asteroseismic analysis (for more details see Zwintz et al. 2008). Good model fits were obtained to the frequencies for $l = 0$, 1 and 2 p modes, but all models lie well outside the observationally derived HR diagram uncertainty box of HD 142666. The reason for this could be that the published value for HD 142666's luminosity is overestimated and/or that additional physics would have to be included in the models to find a good fit to the observed frequencies and the star's position in the HR-diagram (Zwintz et al. 2008).

Acknowledgments. K.Z. and T.K. acknowledge support from the Austrian Science Funds (FWF, KZ: project T335-N16, TK: project P17580). The Natural Sciences and Engineering Research Council of Canada supports the research of D.B.G.

References

Kurtz, D. W., & Müller, M. 2001, MNRAS, 325, 1341

Marconi, M., & Palla, F. 1998, AJ, 507, L141

Meeus, G., Waelkens, C., & Malfait, K. 1998, A&A, 329, 131

Suran, M., Goupil, M., Baglin, A., et al. 2001, A&A, 372, 233

Walker, G., Matthews, J. M., Kuschnig, R., et al. 2003, PASP, 115, 1023

Zwintz, K. 2008, ApJ, 673, 1088

Zwintz, K., Kallinger, T., Guenther, D. B., et al. 2009, A&A, 494, 1031

Comm. in Asteroseismology,
Vol. 159, 2009, JENAM 2008 Symposium № 4: Asteroseismology and Stellar Evolution
S. Schuh & G. Handler

Asteroseismology of chemically peculiar stars

O. Kochukhov

Department of Physics and Astronomy, Uppsala University, Box 515, SE-751 20 Uppsala, Sweden

The section "Rapidly oscillating magnetic Ap stars" of this review is an updated version of the paper published in the proceedings of the Wrocław HELAS Workshop "Interpretation of Asteroseismic Data", CoAst, 157, 228, (Kochukhov 2008).

Abstract

Pulsational variability is observed in several types of main sequence stars with anomalous chemical abundances. In this contribution I summarize the relationship between pulsations and chemical peculiarities, giving special emphasis to rapid oscillations in magnetic Ap stars. These magneto-acoustic pulsators provide unique opportunities to study the interaction of pulsations, chemical inhomogeneities, and strong magnetic fields. Time-series monitoring of rapidly oscillating Ap stars using high-resolution spectrometers at large telescopes and ultra-precise space photometry has led to a number of important breakthroughs in our understanding of these interesting objects. Interpretation of the roAp frequency spectra has allowed constraining fundamental stellar parameters and probing poorly known properties of the stellar interiors. At the same time, investigation of the pulsational wave propagation in chemically stratified atmospheres of roAp stars has been used as a novel asteroseismic tool to study pulsations as a function of atmospheric height and to map in detail the horizontal structure of the magnetically-distorted p modes.

Individual Objects: HR 8799, HD 116114, HD 201601 (γ Equ), HD 176232 (10 Aql), HD 134214, HD 137949 (33 Lib), HD 99563, HD 24712, HD 75445, HD 137909 (β CrB), HD 101065, HR 3831

Introduction

On and near the main sequence, for spectral types from B to early F, one finds a remarkable diversity of the stellar surface properties and variability. In cooler and hotter parts of the H-R diagram a single, powerful process, such as convection in solar-type stars or mass loss in hot massive stars, dominates the physics of stellar atmospheres. In contrast, several processes of comparable magnitude compete in the A-star atmospheres and envelopes, creating interesting and heterogeneous stellar populations. Radiative diffusion (Michaud 1970) is the most important process responsible for non-solar surface chemical composition. Diffusion theory suggests that ions heavier than hydrogen are able to levitate or sink under competing influence of radiation pressure and gravity. Element segregation by radiative diffusion is easily wiped out by various hydrodynamical mixing effects and, thus, requires a star which is stable over a significant part of its outer envelope. Slowly rotating B-F stars with shallow convection zones provide the required stability. The presence of strong, global magnetic fields contributes further to the suppression of turbulence and leads to different diffusion velocities depending on the field inclination and strength. Chemically peculiar stars are separated into two distinct sequences according to their magnetic properties. Am, λ Boo, HgMn stars lack strong magnetic fields and show mild chemical anomalies in chemically homogeneous outer stellar layers. Ap and Bp stars have magnetic fields exceeding few hundred gauss, exhibit extreme chemical anomalies and have substantial vertical and horizontal chemical gradients in the photosphere.

Stellar variability, including pulsations, adds important time-dependent aspects to the complex picture of chemically peculiar B–F stars. Depending on the pulsation frequency and the physics of the interaction between mode excitation, composition gradients and magnetic field, different types of pulsations are suppressed or excited. Observation and asteroseismic interpretation of this pulsational variability is a powerful tool for determining fundamental stellar parameters and constraining poorly known interior properties of chemically peculiar stars.

In this review I summarize our current understanding of the relationship between stellar pulsations and chemical peculiarity for stars in the roAp, δ Scuti, SPB and β Cephei instability regions.

Metallic line A stars

There is near-exclusion of δ Scuti pulsations and Am-type chemical peculiarity (e.g., Breger 1970). Diffusion theory explains this by invoking gravitational settling, which removes He from the He II ionization zone, suppressing the driving of δ Scuti pulsations. When the Am star evolves off the main sequence, the He II ionization region shifts deeper into the star and reaches layers where some residual He is left. This allows excitation of low-amplitude δ Scuti pulsations (Cox et al. 1979). These evolved δ Scuti variables with residual Am-like chemical peculiarities are known as ρ Pup stars.

Classical δ Scuti pulsations have also been claimed in several un-evolved Am and Ap stars (e.g., Kurtz 1989; Martinez et al. 1999; González et al. 2008). In some of these cases the detection of pulsational variability is convincing. However, the Ap or Am nature of the stars in question, inferred from photometry and old low-resolution classification spectra, is very uncertain. Detailed model atmosphere and abundance analyses using modern, high-quality spectroscopic material are required to confirm or refute the suspected unusual combination of high-amplitude δ Scuti pulsation and large chemical peculiarities.

Pulsating λ Bootis stars

λ Boo stars are Population I early-A to early-F type stars which exhibit significant under-abundances of most iron-peak and heavy elements but show solar abundances of CNO and some other light elements (Paunzen et al. 2002a; Heiter 2002). These chemical properties are believed to arise from contamination of the shallow stellar surface convection zones by the accretion of metal-depleted gas from a circumstellar shell (Venn & Lambert 1990) or a diffuse interstellar cloud (Kamp & Paunzen 2002).

The H-R diagram position of the λ Boo group members partially overlaps with the δ Scuti instability strip. While accumulation of metals and He depletion prevents δ Scuti type pulsations in most Ap and Am stars, the opposite abundance signatures of λ Boo stars make them more promising targets for pulsational observations. In particular, asteroseismic investigations of these stars are interesting for constraining the stellar fundamental parameters and determining the average metal content of the stellar interiors.

High-resolution time-series spectroscopy by Bohlender et al. (1999) revealed the presence of high-degree non-radial pulsations in the majority of investigated λ Boo stars. The overall pulsational characteristics of the group were summarized by Paunzen et al. (2002b). They concluded that the fraction of pulsating λ Boo stars inside the δ Scuti instability strip (at least 70%) is significantly larger than for normal stars. Moreover, in contrast to classical δ Scuti stars, which often pulsate in the fundamental mode, λ Boo stars tend to pulsate in high-overtone modes.

Interestingly, at least one object with λ Boo chemical characteristics – the planetary host star HR 8799 – is known to exhibit γ Dor type pulsational variability (Zerbi et al. 1999; Gray & Kaye 1999). However, HR 8799 appears to be an exception as other members of the γ Dor group show normal abundance patterns (Bruntt et al. 2008).

Pulsations and chemical peculiarity in hot stars

The hot pulsating stars (Slowly Pulsating B and β Cephei), chemically peculiar Bp stars, and non-pulsating normal B stars coexist in the same part of the H-R diagram (Briquet et al. 2007). Nevertheless, up to now no conclusive evidence for significant overlap of the pulsational, magnetic, and chemical peculiarity phenomena has been identified for B-type stars. Analysis of the low-resolution UV flux distributions showed that the metallicities of SPB (Niemczura 2003) and β Cephei (Niemczura & Daszynska-Daszkiewicz 2005) pulsators do not differ from those of normal B stars. On the other hand, Morel et al. (2008) suggested the existence of a population of nitrogen-rich and boron-depleted slowly rotating B stars based on NLTE abundance analysis of high-resolution spectra. It is possible that the photospheric chemistry of these objects is altered by a weak magnetic field in qualitatively the same way as much stronger fields of Bp stars lead to prominent deviations from the solar chemical composition. However, apart from a small number of SPB and β Cephei stars with ~ 100 G fields (Neiner et al. 2003; Hubrig et al. 2006a), the universal presence of weak magnetic fields could not be convincingly established for normal and/or pulsating B-type stars.

The non-magnetic HgMn chemically peculiar stars present another challenge for our understanding of the excitation of pulsations in hot stars. Many HgMn stars are situated within the SPB instability strip. Furthermore, an increased opacity due to accumulation of metals by radiative diffusion in HgMn stars is expected to enhance the driving of the SPB pulsations (Turcotte & Richard 2002). Contrary to this theoretical prediction photometric observations show no evidence of pulsational variability in HgMn stars (Adelman 1998). Spectroscopic line profile variations detected for a handful of HgMn stars is limited to lines of 2–3 heavy elements and is, consequently, attributed to chemical inhomogeneities rather than pulsation (Adelman et al. 2002; Kochukhov et al. 2005; Hubrig et al. 2006b). Incompleteness of the theoretical diffusion models in the outer part of the stellar envelope is the most likely explanation for the contradiction between predicted and observed pulsation properties of HgMn stars.

Rapidly oscillating magnetic Ap stars

Rapidly oscillating Ap (roAp) stars represent the most prominent subgroup of pulsating chemically peculiar stars. These objects belong to the SrCrEu type of magnetic A stars, and pulsate in high-overtone, low-degree p modes. roAp stars are found at or near the main sequence, at the cool border of the region occupied by the magnetic Ap/Bp stars (Kochukhov & Bagnulo 2006). According to a series of recent spectroscopic studies (e.g., Ryabchikova et al. 2002, 2004; Kochukhov et al. 2002a), the effective temperatures of roAp stars range from about 8100 down to 6400 K. Their atmospheres are characterized by diverse chemical abundance patterns, but typically have normal or below solar concentration of light and iron-peak elements and a very large overabundance of rare-earth elements (REEs). Similar to other cool magnetic A stars, roAp stars possess global fields with a typical strength from a few to ten kG (Mathys et al. 1997), although in some stars the field intensity can exceed 20 kG (Kurtz et al. 2006b). These global magnetic topologies are most likely the remnants of the fields which were swept at the star-formation phase or generated by a dynamo in the convective envelopes of pre-main sequence stars, then quickly decayed to a stable configuration (Braithwaite & Nord-lund 2006) and now remain nearly constant on stellar evolutionary time scales. The slow rotation and stabilizing effect of the strong magnetic field facilitates the operation of atomic diffusion processes (Michaud et al. 1981; LeBlanc & Monin 2004), which are responsible for a grossly non-solar surface chemistry and large element concentration gradients in Ap star atmospheres (Ryabchikova et al. 2002, 2008; Kochukhov et al. 2006). Variation of the field strength and inclination across the stellar surface alters the local diffusion velocities (Alecian & Stift 2006), leading to the formation of spotted chemical distributions and consequential synchronous rotational modulation of the broad-band photometric indices, spectral line profiles, the longitudinal magnetic field and the mean field modulus (e.g., Ryabchikova et al. 1997).

Pulsations in cool Ap stars were discovered 30 years ago (Kurtz 1978) and were immediately recognized to be another manifestation of the prominent influence of unusually strong magnetic fields on the stellar interiors and atmospheres. Currently, 38 cool Ap stars are known to pulsate. Several new roAp stars were recently discovered using high-resolution spectroscopic observations (Hatzes & Mkrtichian 2004; Elkin et al. 2005; Kurtz et al. 2006b; Kochukhov et al. 2008a, 2009; González et al. 2008). Oscillations have amplitudes below 10 mmag in the Johnson B filter and 0.05–5 $km\,s^{-1}$ in spectroscopy, while the periods lie in the range from 4 to 22 min. The latter upper period threshold of roAp pulsation corresponds to the second mode recently detected by the high-precision RV observations of the evolved Ap star HD 116114 (Kochukhov, Bagnulo & Lo Curto, in preparation).

The amplitude and phase of pulsational variability are modulated with the stellar rotation. A simple geometrical interpretation of this phenomenon was suggested by the oblique pulsator model of Kurtz (1982), which supposes an alignment of the low angular degree modes with the quasi-dipolar magnetic field of the star and resulting variation of the aspect at which pulsations are seen by the distant observer. Detailed theoretical studies (Bigot & Dziembowski 2002; Saio 2005) showed that the horizontal geometry of p-mode pulsations in magnetic stars is far more complicated: individual modes are distorted by the magnetic field and rotation in such a way that pulsational perturbation cannot be approximated by a single spherical harmonic function.

Photometric studies of roAp pulsations

Most roAp stars were discovered by D. Kurtz and collaborators using photometric observations at SAAO (see review by Kurtz & Martinez 2000). The search for roAp stars in the northern hemisphere is being conducted at the Nainital (Joshi et al. 2006) and Maidanak (Dorokhova & Dorokhov 2005) observatories. Several roAp stars were observed in coordinated multisite photometric campaigns (Kurtz et al. 2005a; Handler et al. 2006), which allowed to deduce frequencies with a precision sufficient for asteroseismic analysis. However, the low amplitudes of broad-band photometric variation of roAp stars, low duty cycles and aliasing problems inevitably limit the precision of ground-based photometry. Instead of pursuing observations from the ground, recent significant progress has been achieved by uninterrupted, ultra-high precision observations of known roAp stars using small photometric telescopes in space. Here the Canadian MOST space telescope is the undisputed leader. The MOST team has completed 3–4 week runs on HD 24712, γ Equ (HD 201601), 10 Aql (HD 176232), HD 134214, and HD 99563. Observations of 33 Lib (HD 137949) are planned for April-May 2009.

In addition to providing unique material for detailed asteroseismic studies of HD 24712, γ Equ, and 10 Aql, the MOST photometry has revealed the presence of a very close frequency pair in γ Equ, giving a modulation of pulsation amplitude with a ≈18 d period (Huber et al. 2008). It is possible that this frequency beating is responsible for the puzzling discrepancy of the radial velocity amplitudes found for γ Equ in different short spectroscopic observing runs (Sachkov et al. 2008). This amplitude variation could not be ascribed to the rotational modulation because the rotation period of this star exceeds 70 years (Bychkov et al. 2006).

Spectroscopy of roAp pulsations

High-quality time-resolved spectra of roAp stars have proven to be the source of new, incredibly rich information, which not only opened new possibilities for the research on magneto-acoustic pulsations but yielded results of wide astrophysical significance. Numerous spectroscopic studies of individual roAp stars (e.g., Kochukhov & Ryabchikova 2001a; Mkrtichian et al. 2003; Ryabchikova et al. 2007a), as well as a comprehensive analysis of pulsational variability in 10 roAp stars published by Ryabchikova et al. (2007b), demonstrated pulsations in spectral lines very different from those observed in any other type of non-radially pulsating stars. The most prominent characteristic of the RV oscillation in roAp stars is the extreme diversity of pulsation signatures of different elements. Only a few stars show evidence of <50 m s^{-1} variation in the lines of iron-peak elements, whereas REE lines, especially those of Nd II, Nd III, Pr III, Dy III, and Tb III exhibit amplitudes ranging from a few hundred m s^{-1} to several km s^{-1}. The narrow core of Hα behaves similarly to REE lines (Kochukhov 2003; Ryabchikova et al. 2007b), suggesting line formation at comparable atmospheric heights.

The pulsation phase also changes significantly from one line to another (Kochukhov & Ryabchikova 2001a; Mkrtichian et al. 2003), with the most notorious example of 33 Lib where different lines of *the same ion* pulsate with a 180° shift in phase, revealing a radial node, and show very different ratios of the amplitude at the main frequency and its first harmonic (Ryabchikova et al. 2007b). Several studies concluded that, in general, roAp stars show a combination of running (changing phase) and standing (constant phase) pulsation waves at different atmospheric heights.

Another unusual aspect of the spectroscopic pulsations in roAp stars is a large change of the oscillation amplitude and phase from the line core to the wings. The bisector variation expected for the regular spherical harmonic oscillation is unremarkable and should exhibit neither changing phase nor significantly varying amplitude. Contrary to this expectation of the common single-layer pulsation model, the roAp bisector amplitude often shows an increase from 200–400 m s^{-1} in the cores of strong REE lines to 2–3 km s^{-1} in the line wings, accompanied by significant changes of the bisector phase (Sachkov et al. 2004; Kurtz et al. 2005b; Ryabchikova et al. 2007b).

The ability to resolve and measure with high precision pulsational variation in individual lines allows to focus the analysis on the spectral features most sensitive to pulsations. By co-adding radial velocity curves of many REE lines recorded in a spectrum with a wide wavelength coverage one is able to reach a RV accuracy of ~ 1 m s^{-1} (Mathys et al. 2007). This made possible the discovery of very low-amplitude oscillations in HD 75445 (Kochukhov et al. 2009) and HD 137909 (Hatzes & Mkrtichian 2004). The second object, the well-known cool Ap star β CrB, was previously considered to be a typical non-oscillating Ap (noAp) star due to null results of numerous photometric searches of pulsations (Martinez & Kurtz 1994) and the absence of prominent REE ionization anomaly found for nearly all other roAp stars (Ryabchikova et al. 2001, 2004). The fact that β CrB is revealed as the second brightest roAp star corroborates the idea that p mode oscillations could be present in all cool Ap stars but low pulsation amplitudes prevented detection of pulsations in the so-called noAp stars (Kochukhov et al. 2002b; Ryabchikova et al. 2004).

Despite the improved sensitivity in searches of the low-amplitude oscillations in roAp candidates and numerous outstanding discoveries for known roAp stars, the major limitation of the high-resolution spectroscopic monitoring is a relatively small amount of observing time available at large telescopes for these projects. As a result, only snapshot time-series spanning 2–4 hours were recorded for most roAp stars, thus providing an incomplete and possibly biased picture for the multiperiodic pulsators, for which close frequencies cannot be resolved in such short runs. Observations on different nights, required to infer detailed RV frequency spectra, were secured only for a few roAp stars (Kochukhov 2006; Mkrtichian et al. 2008). For example, in a recent multisite spectroscopic campaign carried out for 10 Aql using two telescopes on seven observing nights (Sachkov et al. 2008), we found that beating of the three dominant frequencies leads to strong changes of the apparent RV amplitude during several hours. This phenomenon could explain the puzzling modulation of the RV pulsations on time scales of 1–2 hours detected in some roAp stars (Kochukhov & Ryabchikova 2001b; Kurtz et al. 2006a).

Asteroseismology of roAp stars

The question of the roAp excitation mechanism has been debated for many years but now is narrowed down to the κ mechanism acting in the hydrogen ionization zone, with the additional influence from the magnetic quenching of convection and composition gradients built up by atomic diffusion (Balmforth et al. 2001; Cunha 2002; Vauclair & Théado 2004). However, theories cannot reproduce the observed temperature and luminosity distribution of roAp stars and have not been able to identify parameters distinguishing pulsating Ap stars from their apparently constant, but otherwise very similar, counterparts (Théado et al. 2009). At the same time, some success has been achieved in calculating the magnetic perturbation of oscillation frequencies (Cunha & Gough 2000; Saio & Gautschy 2004) and inferring fundamental parameters and interior properties for multiperiodic roAp stars (Matthews et al. 1999; Cunha et al. 2003).

The recent asteroseismic interpretation of the frequencies deduced from the MOST data for γ Equ (Gruberbauer et al. 2008) and 10 Aql (Huber et al. 2008) yields stellar parameters in good agreement with detailed model atmosphere studies. At the same time, the magnetic field required by seismic models to fit the observed frequencies is 2–3 times stronger than the field modulus inferred from the Zeeman split spectral lines. This discrepancy could be an indication that the magnetic field in the p-mode driving zone is significantly stronger than the surface field or it may reflect an incompleteness of the theoretical models.

Mkrtichian et al. (2008) presented the first detailed asteroseismic analysis of a roAp star based entirely on spectroscopic observations. Using high-precision RV measurements spanning four consecutive nights, the authors detected 26 frequencies for the famous roAp star HD 101065 (Przybylski's star). Mode identification showed the presence of 15 individual modes with $l = 0$–2. This rich frequency spectrum of HD 101065 can be well reproduced by theoretical models if an excessively strong ($\approx 9\,\mathrm{kG}$) dipolar magnetic field is assumed, in contradiction to $\langle B \rangle = 2.3\,\mathrm{kG}$ inferred directly from the stellar spectrum (Cowley et al. 2000).

Tomography of atmospheric pulsations in roAp stars

The key observational signature of roAp pulsations in spectroscopy – large line-to-line variation of pulsation amplitude and phase – is understood in terms of an interplay between pulsations and chemical stratification. The studies by Ryabchikova et al. (2002, 2008) and Kochukhov et al. (2006) demonstrated that light and iron-peak elements tend to be overabundant in deep atmospheric layers (typically $\log \tau_{5000} \geq -0.5$) of cool Ap stars, which agrees with the predictions of self-consistent diffusion models (LeBlanc & Monin 2004). On the other hand, REEs accumulate in a cloud at very low optical depth. The NLTE stratification studies, performed for Nd and Pr ions, place the lower boundary of this cloud at $\log \tau_{5000} \approx -3$ (Mashonkina et al. 2005, 2009). Then, the rise of pulsation amplitude towards the upper atmospheric layers due to the exponential density decrease does not affect Ca, Fe, and Cr lines but shows up prominently in the core of Hα and in REE lines. This picture of the pulsation waves propagating outwards through the stellar atmosphere with highly inhomogeneous chemistry has gained general support from observations and theoretical studies alike. Hence the properties of roAp atmospheres allow an entirely new type of asteroseismic analysis – vertical resolution of p mode cross-sections simultaneously with constraints on the distribution of chemical abundances.

The two complimentary approaches to the roAp pulsation tomography problem have been discussed by Ryabchikova et al. (2007a, 2007b). On the one hand, tedious and detailed line formation calculations, including stratification analysis, NLTE line formation, sophisticated model atmospheres and polarized radiative transfer, can supply mean formation heights for individual pulsating lines. Then, the pulsation mode structure can be mapped directly by plotting pulsation amplitudes and phases of selected lines against optical or geometrical depth. On the other hand, the phase-amplitude diagram method proposed by Ryabchikova et al. (2007b) is suitable for a coarse analysis of the vertical pulsation structure without invoking model atmosphere calculations but assuming the presence of an outwards propagating wave characterized by a continuous change of amplitude and phase. In this case, a scatter plot of the RV measurements in the phase-amplitude plane can be interpreted in terms of standing and running waves, propagating in different parts of the atmosphere.

To learn about the physics of roAp atmospheric oscillations one should compare empirical pulsation maps with theoretical models of the p mode propagation in the magnetically-dominant ($\beta \ll 1$) part of the stellar envelope. Sousa & Cunha (2008) considered an analytical model of radial modes in an isothermal atmosphere with exponential density decrease. They argue that waves are decoupled into the standing magnetic and running acoustic components, oriented perpendicular and along the magnetic field lines, respectively. The total projected pulsation velocity, produced by a superposition of these two components, can have a widely different vertical profile depending on the magnetic field strength, inclination and the aspect angle. For certain magnetic field parameters and viewing geometries the two components cancel out, creating a node-like structure. This model can possibly account for observations of radial nodes in 33 Lib (Mkrtichian et al. 2003) and 10 Aql (Sachkov et al. 2008).

The question of interpreting the line profile variations (LPV) of roAp stars has received great attention after it was demonstrated that the REE lines in γ Equ exhibit an unusual blue-to-red asymmetric variation (Kochukhov & Ryabchikova 2001a), which is entirely unexpected for a slowly rotating non-radial pulsator. Kochukhov et al. (2007) showed the presence of similar LPV in the REE lines of several other roAp stars and presented examples of the transformation from the usual symmetric blue-red-blue LPV in Nd II lines to the asymmetric blue-to-red waves in the Pr III and Dy III lines formed higher in the atmosphere. These lines also show anomalously broad profiles (e.g., Ryabchikova et al. 2007b), suggesting the existence of an isotropic velocity field, with a dispersion of the order of 10 km s^{-1}, in the uppermost atmospheric layers. Kochukhov et al. (2007) proposed a phenomenological model of the interaction between this turbulent layer and pulsations that has successfully reproduced asymmetric LPV of doubly ionized REE lines. An alternative model by Shibahashi et al. (2008) obtains similar LPV by postulating formation of REE lines at extremely low optical depths, in disagreement with the detailed NLTE calculations by Mashonkina et al. (2005, 2009), and requires the presence of shock waves in stellar atmospheres, which is impossible to reconcile with the fact that observed RV amplitudes are well below the sound speed.

Oblique pulsations and distortion of non-radial modes by rotation and magnetic fields preclude direct application of the standard mode identification techniques to roAp stars. A meaningful study of their horizontal pulsation geometry became possible by using the method of pulsation Doppler imaging (Kochukhov 2004a). This technique derives maps of pulsational fluctuations without making a priori assumptions of the spherical harmonic pulsation geometry. Application of this method to HR 3831 (Kochukhov 2004b) provided the first independent verification of the oblique pulsator model by showing alignment of the axisymmetric pulsations with the symmetry axis of the stellar magnetic field. At the same time, Saio (2005) showed that the observed deviation of the oscillation geometry of HR 3831 from a oblique dipole mode agrees well with his model of magnetically distorted pulsation.

Outlook

A progress in understanding the relation between the phenomena of chemical peculiarity and stellar pulsations calls for a detailed model atmosphere and chemical abundance analysis of the suspected high-amplitude δ Scuti Ap and classical Am stars. Interpretation of the modern high-resolution spectroscopic observational material using realistic model atmospheres is also needed to clarify the question of the connection between CP and hot pulsating stars. Systematic high-resolution spectropolarimetric observations are urgently needed to verify the claims of weak magnetic fields in many SPB and a few β Cephei stars. On the other hand, comprehensive theoretical modelling is needed to explore the asteroseismic potential of the pulsating λ Boo stars and, in particular, to test possibilities of constraining their interior chemical profiles.

For roAp stars, several important open questions and promising research directions can be identified. On the theoretical side, the failure of the current pulsation models to account for the observed blue and red borders of the roAp instability strip should be addressed by including a more realistic physical description of the interplay between pulsations, magnetic fields, stratified chemistry, and stellar rotation. On the observational side, systematic spectroscopic searches for low-amplitude magneto-acoustic oscillations in cool Ap stars are evidently needed to overcome the limitations and biases of previous photometric surveys.

The remarkable spectroscopic pulsational behaviour, demonstrated in numerous recent studies of roAp stars, extends the roAp research to the uncharted territory far beyond the field of classical asteroseismology. In addition to interpretation of pulsation frequencies, roAp stars now offer a unique opportunity for *pulsation tomography*, i.e. a study of different pulsation modes in 3-D, made possible by the rotational modulation of the oblique pulsations and the prominent effect of chemical stratification. Spectacular observational results, such as resolution of the vertical pulsation mode cross-sections and Doppler imaging of atmospheric pulsations in roAp stars, are, however, yet to be matched by corresponding theoretical developments. At the moment we lack realistic models treating propagation of pulsation waves in the outer layers of magnetic Ap stars. Our knowledge about chemical stratification, in particular that of rare-earth elements, and its impact on the atmospheric structure is equally incomplete. Addressing these theoretical questions is required for the development of a solid physical basis for astrophysical interpretation of the recent roAp pulsation tomography results.

References

Adelman, S. J. 1998, A&ASS, 132, 93

Adelman, S. J., Gulliver, A. F., Kochukhov, O. P., & Ryabchikova, T. A. 2002, ApJ, 575, 449

Alecian, G., & Stift, M. J. 2006, A&A, 454, 571

Balmforth, N. J., Cunha, M. S., Dolez, N., et al. 2001, MNRAS, 323, 362

Bigot, L., & Dziembowski, W. A. 2002, A&A, 391, 235

Bohlender, D. A., González, J. F., & Matthews, J. M. 1999, A&A, 350, 553

Braithwaite, J., & Nordlund, Å. 2006, A&A, 450, 1077

Breger, M. 1970, ApJ, 162, 597

Briquet, M., Hubrig, S., De Cat, P., et al. 2007, A&A, 466, 269

Bruntt, H., De Cat, P., & Aerts, C. 2008, A&A, 478, 487

Bychkov, V. D., Bychkova, L. V., & Madej, J. 2006, MNRAS, 365, 585

Cowley, C. R., Ryabchikova, T., Kupka, F., et al. 2000, MNRAS, 317, 299

Cox, A. N., King, D. S., & Hodson, S. W. 1979, ApJ, 231, 798

Cunha, M. S., & Gough, D. 2000, MNRAS, 319, 1020

Cunha, M. S. 2002, MNRAS, 333, 47

Cunha, M. S., Fernandes, J. M. M. B., & Monteiro, M. J. P. F. G. 2003, MNRAS, 343, 831

Dorokhova, T., & Dorokhov, N. 2005, JApA, 26 223

Elkin, V. G., Rilej, J., Cunha, M. S., et al. 2005, MNRAS, 358, 665

González, J. F., Hubrig, S., Kurtz, D. W., Elkin, V., & Savanov, I. 2008, MNRAS, 384, 1140

Gruberbauer, M., Saio, H., Huber, D., et al. 2008, A&A, 480, 223

Gray, R.O, & Kaye, A. B. 1999, AJ, 118, 2993

Handler, G., Weiss, W. W., Shobbrook, R. R., et al. 2006, MNRAS, 366, 257

Hatzes, A. P., & Mkrtichian, D. E. 2004, MNRAS, 351, 663

Heiter, U. 2002, A&A, 381, 959

Huber, D., Saio, H., Gruberbauer, M., et al. 2008, A&A, 483, 239

Hubrig, S., Briquet, M., Schöller, M., et al. 2006a, MNRAS, 369, L61

Hubrig, S., González, J. F., Savanov, I., et al. 2006b, MNRAS, 371, 1953

Joshi, S., Mary, D. L., Martinez, P., et al. 2006, A&A, 455, 303

Kamp, I., & Paunzen, E. 2002, MNRAS, 335, L45

Kochukhov, O., & Ryabchikova, T. 2001a, A&A, 374, 615

Kochukhov, O., & Ryabchikova, T. 2001b, A&A, 377, L22

Kochukhov, O., Landstreet, J. D., Ryabchikova, T., Weiss, W. W., & Kupka, F. 2002b, MNRAS, 337, L1

Kochukhov, O. 2003, in "Magnetic Fields in O, B and A stars", eds. L. A. Balona, H. F. Henrichs, and R. Medupe, ASP Conf. Ser., 305, 104

Kochukhov, O. 2004a, A&A, 423, 613

Kochukhov, O. 2004b, ApJ, 615, L149

Kochukhov, O., Piskunov, N., Sachkov, M., & Kudryavtsev, D. 2005, A&A, 439, 1093

Kochukhov, O. 2006, A&A, 446, 1051

Kochukhov, O. 2008, CoAst, 157, 228

Kochukhov, O., & Bagnulo, S. 2006, A&A, 450, 763

Kochukhov, O., Bagnulo, S., & Barklem, P.S. 2002, ApJ, 578, L75

Kochukhov, O., Tsymbal, V., Ryabchikova, T., et al. 2006, A&A, 460, 831

Kochukhov, O., Ryabchikova, T., Weiss, W. W., et al. 2007, MNRAS, 376, 651

Kochukhov, O., Ryabchikova, T., Bagnulo, S., & Lo Curto, G. 2008, A&A, 479, L29

Kochukhov, O., Bagnulo, S., Lo Curto, G., & Ryabchikova, T., 2009, A&A, 493, 45

Kurtz, D. W. 1978, IBVS, 1436

Kurtz, D. W. 1982, MNRAS, 200, 807

Kurtz, D. W. 1989, MNRAS, 238, 1077

Kurtz, D. W., & Martinez, P. 2000, Baltic Astronomy, 9, 253

Kurtz, D. W., Elkin, V.G., & Mathys, G. 2005a, MNRAS, 358, L10

Kurtz, D. W., Cameron, C., Cunha, M. S., et al. 2005b, MNRAS, 358, 651

Kurtz, D. W., Elkin, V.G., & Mathys, G. 2006a, MNRAS, 370, 1274

Kurtz, D. W., Elkin, V.G., Cunha, M. S., et al. 2006b, MNRAS, 372, 286

LeBlanc, F., & Monin, D. 2004, in "The A-Star Puzzle", eds. J. Zverko, J. Ziznovsky, S. J. Adelman, and W. W. Weiss, IAU Symposium, 224, 193

Mashonkina, L., Ryabchikova, T., & Ryabtsev, V. 2005, A&A, 441, 309

Mashonkina, L., Ryabchikova, T., Ryabtsev, A., & Kildiyarova, R. 2009, A&A, 495, 297

Mathys, G., Hubrig, S., Landstreet, J. D., et al. 1997, A&AS, 123, 353

Mathys, G., Kurtz, D. W., & Elkin, V. G. 2007, MNRAS, 380, 181

Matthews, J. M, Kurtz, D. W., & Martinez, P. 1999, ApJ, 511, 422

Martinez, P., & Kurtz, D. W. 1994, MNRAS, 271, 129

Martinez, P., Kurtz, D. W., & Ashoka, B. N., et al. 1999, MNRAS, 309, 871

Michaud, G. 1970, ApJ, 160, 641

Michaud, G., Charland, Y., & Megessier, C. 1981, A&A, 103, 244

Mkrtichian, D. E., Hatzes, A. P., & Kanaan, A. 2003, MNRAS, 345, 781

Mkrtichian, D. E., Hatzes, A. P., Saio, H., & Shobbrook, R. R. 2008, A&A, 490, 1109

Morel, T., Hubrig, S., & Briquet, M. 2008, A&A, 481, 453

Neiner, C., Geers, V. C., Henrichs, H. F., et al. 2003, A&A, 406, 1019

Niemczura, E. 2003, A&A, 404, 689

Niemczura, E., & Daszynska-Daszkiewicz, J. 2005, A&A, 433, 659

Paunzen, E., Iliev, I. Kh., Kamp, I., & Barzova, I. S. 2002a, MNRAS, 336, 1030

Paunzen, E., Handler, G., Weiss, W. W., et al. 2002b, A&A, 392, 515

Ryabchikova, T. A., Landstreet, J. D., Gelbmann, M. J., et al. 1997, A&A, 327, 1137

Ryabchikova, T. A., Savanov, I. S., Malanushenko, V. P., & Kudryavtsev, D. O. 2001, Astron. Reports, 45, 382

Ryabchikova, T., Piskunov, N., Kochukhov, O., et al. 2002, A&A, 384, 545

Ryabchikova, T., Nesvacil, N., Weiss, W. W., et al. 2004, A&A, 423, 705

Ryabchikova, T., Sachkov, M., Weiss, W. W., et al. 2007a, A&A, 462, 1103

Ryabchikova, T., Sachkov, M., Kochukhov, O., & Lyashko, D. 2007b, A&A, 473, 907

Ryabchikova, T., Kochukhov, O., & Bagnulo, S. 2008, A&A, 480, 811

Sachkov, M., Ryabchikova, T., Kochukhov, O., et al. 2004, in "IAU Colloquium 193: Variable Stars in the Local Group", eds. D. W. Kurtz and K. R. Pollard, ASP Conf. Ser., 310, 208

Sachkov, M., Kochukhov, O., Ryabchikova, T., et al. 2008, MNRAS, 389, 903

Sachkov, M., Ryabchikova, T., Gruberbauer, M., & Kochukhov, O. 2008, in "Interpretation of Asteroseismic Data", CoAst, 157, 363

Saio, H., & Gautschy, A. 2004, MNRAS, 350, 485

Saio, H. 2005, MNRAS, 360, 1022

Shibahashi, H., Gough, D., Kurtz, D. W., & Kambe, E. 2008, PASJ, 60, 63

Sousa, J., & Cunha, M. S. 2008, CoSka, 38, 453

Théado, S., Dupret, M.-A., Noels, A., & Ferguson, J.W. 2009, A&A, 493, 159

Turcotte, S., & Richard, O. 2002, Ap&SS, 284, 225

Vauclair, S., & Théado, S. 2004, A&A, 425, 179

Venn, K. A., & Lambert, D. L. 1990, ApJ, 363, 234

Zerbi, F. M., Rodriguez, E., Garrido, R., et al. 1999, MNRAS, 303, 275

Comm. in Asteroseismology,
Vol. 159, 2009, JENAM 2008 Symposium № 4: Asteroseismology and Stellar Evolution
S. Schuh & G. Handler

The pulsating component of the B[e]/X-Ray transient and multiple system
CI Cam (XTE J0421+560)

E. A. Barsukova [1] and V. P. Goranskij [2]

[1] Special Astrophysical Observatory, Nizhny Arkhyz, Karachai-Cherkesia, 369167 Russia
[2] Sternberg Astronomical Institute, Moscow University, Moscow, 119992 Russia

Abstract

We provide evidence that the rapid intranight variability of CI Cam is due to multiperiodic pulsations of the primary B4III-V component.

Individual Objects: CI Cam

CI Cam underwent an outburst in all ranges of electromagnetic waves in 1998. This is a multiple system which consists of a B4 III-V star displaying the B[e] phenomenon, a compact object, probably a white dwarf (WD) on an eccentric orbit with the period of 19.407 day (Barsukova et al. 2006), and a third massive component of unknown nature which causes the slow shift of the wind lines of the B[e] star including the forbidden line [N II] 5755 Å (Barsukova et al. 2007). The 1998 outburst is treated by some investigators as a thermonuclear explosion of hydrogen accumulated on the surface of WD from the dense circumstellar envelope and stellar wind. So, CI Cam is a unique system which resembles a classical nova.

We carried out an extensive photometric CCD monitoring in the V band using the SAI Crimean Station 50-cm Maksutov telescope during 18 nights in 2006 December. The set includes 2366 observations. A fragment of the V-band curve is presented in Fig. 1a. The duration of night monitoring reached 13.2 hours in some nights, and the accuracy of measurements was between 2 and 4 mmag. Methods of analysis were discrete Fourier transform with decomposition into periodic components including a prewhitening procedure and light curve model reconstruction. The set was cleaned for low-frequency noise at frequencies less than 0.2 c/d. The amplitude spectrum in the frequency range $0.2-20$ c/d is shown in Fig. 2.

The intranight variability of CI Cam is due to multiperiodic pulsations of its B4-type component (Barsukova & Goranskij 2008). Two waves predominate with periods of 0.4152 and 0.2667 day and full amplitudes of 19 and 17 mmag, correspondingly. They are extracted and shown in Figs. 1c and 1d, separately. The ratio of their periods is close to 3:2. At times, the maximum pulsation amplitude reaches 70 mmag owing to a resonance between these waves, and the pulsations are seen as mini-flares. Pulsations of CI Cam resemble those of Be stars, but they are observed in a B[e] star for the first time. The details of this study will be published in the paper by Goranskij & Barsukova 2009.

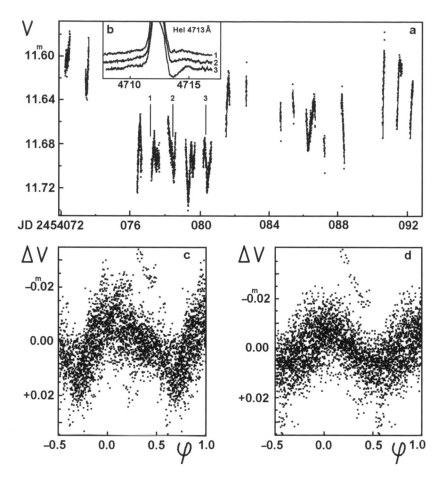

Figure 1: Light curves of CI Cam. a) Fragment of the V-band light curve taken in 2006 December. Digits 1–3 denote times of high resolution spectra using BTA/NES. b) HeI 4713 Å line profiles in the moments of spectral observations. c) Data cleaned for low-frequency noise and periodic variation with $P = 0.2667$ day are folded with $P = 0.4152$ day. d) The same data cleaned for the noise and periodic variation with $P = 0.4152$ day are folded with $P = 0.2667$ day.

We analyzed 20 high-resolution spectra of CI Cam, three of them were taken with the Russian 6-m telescope in the nights of our photometric monitoring. One of these three spectra (No. 3) shows inverse P Cyg type profiles in the weak HeI lines (Fig. 1b). In the total sample of spectra, three show inverse P Cyg profiles in HeI lines and one spectrum displays a classical P Cyg profile. This behaviour is typical for pulsations. The matter thrown out into the stellar envelope by shock waves falls backwards and absorbs photospheric radiation. In this sense, high-resolution spectroscopic observations allow to distinguish pulsations from circumstellar phenomena, while photometric variations potentially not only include photospheric but also circumstellar variability.

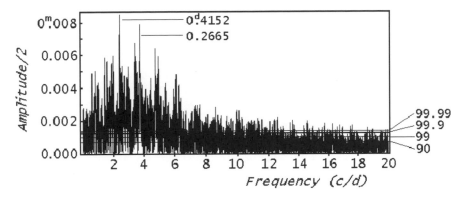

Figure 2: The amplitude spectrum of CI Cam. Horizontal lines and digits to the right are significance levels and their corresponding per cent values.

Evidently, the pulsation of CI Cam, like pulsations in Be stars, is a factor playing an important role in the formation of the circumstellar disk and gas transfer from the star into the disk. Additionally, our result may be an asteroseismologic clue to the further study of the internal structure of stars with the B[e] phenomenon.

Acknowledgments. We thank the Russian Foundation for Basic Research for the Grants No.06-02-16865 and 07-02-00630. E.A.B. is grateful to the Organizing Committee of the Symposium for the invitation.

References

Barsukova, E. A., Borisov, N. V., Burenkov, A. N., et al. 2006, Astronomy Reports, 50, 664

Barsukova, E. A., Klochkova, V. G., Panchuk, V. E., et al. 2007, ATel, 1036

Barsukova, E. A., & Goranskij, V. P. 2008, ATel, 1381

Goranskij, V. P., & Barsukova, E. A. 2009, Astrophysical Bulletin, 64, 50

Comm. in Asteroseismology,
Vol. 159, 2009, JENAM 2008 Symposium № 4: Asteroseismology and Stellar Evolution
S. Schuh & G. Handler

Asteroseismology and evolution of EHB stars

R. H. Østensen

Instituut voor Sterrenkunde, K.U.Leuven, Celestijnenlaan 200D, B-3001 Leuven, Belgium

Abstract

The properties of the Extreme Horizontal Branch stars are quite well understood, but much uncertainty surrounds the many paths that bring a star to this peculiar configuration. Astero-seismology of pulsating EHB stars has been performed on a number of objects, bringing us to the stage where comparisons of the inferred properties with evolutionary models becomes feasible. In this review I will outline our current understanding of the formation and evolution of these stars, with emphasis on recent progress. The aim is to show how the physical param-eters derived by asteroseismology can enable the discrimination between different evolutionary models.

Individual Objects: V361 Hya, V1093 Her, DW Lyn, V391 Peg, Balloon 090100001, KL UMa, NY Vir, V338 Ser, LS IV–14°116, HD 188112

Introduction

Let me first clarify the basic terminology with respect to EHB stars, which can sometimes be confusing as the terms EHB and sdB are often used to label the same stars. The terms Extended Horizontal Branch or Extreme Horizontal Branch have been used interchangeably to describe the sequence of stars observed to lie bluewards of the normal Horizontal Branch stars in globular clusters, and also in temperature/gravity plots of hot field stars. The EHB feature was first described and associated with field sdB and sdO stars by Greenstein & Sargent (1974). Now, EHB stars are taken to mean core helium burning stars with an envelope too thin to sustain hydrogen burning. It is also understood that not all sdB stars are EHB stars. In particular, if a star loses its envelope without the core reaching the mass required for the helium flash, its cooling track can take it through the sdB domain on its way to become a helium core WD. The sdB/sdO terms are used to describe the spectroscopic appearance and do not presume any particular evolutionary stage. Several subclassification schemes have been used, but most common nowadays is the one introduced by Moehler et al. (1990). This scheme names as sdB stars those of the hot subdwarfs showing HeI absorption lines, as sdO stars those showing HeII, and as sdOB stars those showing features of both. Additionally the terms He-sdB and He-sdO are used to describe stars in which the helium lines dominate over the Balmer lines. The EHB forms a sequence of stars from the coolest sdBs to the sdOB domain, and it is clear that most stars given this classification are in fact EHB stars. For the He-rich objects a coherent picture has yet to emerge.

The current canonical picture of the EHB stars was mostly established by Heber (1986), in which the EHB stars are helium core burning stars with masses close to the core helium flash mass of $\sim 0.47\,M_\odot$, and an extremely thin hydrogen envelope, too thin to sustain hydrogen burning (no more than 1% by mass). It is understood that they are post red giant branch

(RGB) stars that have started core helium burning in a helium flash before or after the envelope was removed by any of several possible mechanisms. The lifetime of EHB stars from the zero-age EHB (ZAEHB) to the terminal age EHB (TAEHB), when core helium runs out, takes between 100 and 150 Myrs. The post-EHB evolution will take them through the sdO domain directly to the white dwarf (WD) cooling curve without ever passing through a second giant stage. The time they spend shell helium burning before leaving the sdO domain can be up to 20 Myrs.

Although the future evolution of EHB stars after core He-exhaustion has always been presumed quite simple, the paths that lead to the EHB have always been somewhat mysterious. New hope that the evolutionary paths leading to the formation of EHB stars can be resolved has been kindled by the discovery that many of them pulsate, which has opened up the possibility of probing their interiors using asteroseismological methods. These pulsators are known as sdBV stars, and several distinct subclasses are now recognized (see the *Asteroseismology* section below). But in order to understand what questions asteroseismology can ask and answer, it is essential to understand the different paths that produce EHB stars. Only by understanding the evolutionary history of these stars is it possible to construct realistic models of their interiors which are needed for asteroseismology to be able to distinguish between the different formation scenarios. For this reason we will review the essential points of the *Formation and Evolution* first, after starting with a look at the observed properties of the hot subdwarf population in *The Observed EHB* below.

Besides the spectacularly rapid pulsations in the EHB stars, another factor that has contributed to the recent burst in interest in EHB stars is the realization that these stars are the main contributor to the UV-upturn phenomenon observed in elliptical galaxies. An excellent review of the UV upturn and the binary population synthesis models required to model this phenomenon can be found in Podsiadlowski et al. (2008). For a more in-depth review of the properties of all hot subdwarf stars, the exhaustive review by Heber (2009) is recommended.

The Observed EHB

Hot subdwarf stars were found in the galactic caps already by Humason & Zwicky (1947). By the time Greenstein & Sargent (1974) wrote their seminal paper, the number of such faint blue stars had grown to 189, permitting a systemic study of the population. The PG survey (Green et al. 1986), which covered more than 10 000 square degrees at high galactic latitudes, found that of 1874 UV-excess objects detected more than 1000 were hot subdwarfs, so these stars dominate the population of faint blue stars down to the PG survey limit ($B = 16.5$). Together with the large sample of subdwarfs detected in the HS survey and analyzed by Edelmann et al. (2003), these have provided a rich source of hot subdwarfs for observers to follow up, and new discoveries are still being made. The recent Sloan Digital Sky Survey (SDSS, Stoughton et al. 2002) also contains spectra of more than 1000 hot subdwarfs, but as the SDSS reaches much deeper than the PG survey, WD stars start to dominate around about $B = 18$ as the thickness of the galactic disk is reached. The Subdwarf Database (Østensen 2006) catalogs about 2500 hot subdwarfs, with extensive references to the available literature.

Several surveys have attempted to tackle the question of the binary frequency on the EHB, but the matter is complicated by the many different types of systems EHB stars are found in. EHB stars with FGK companions are easily detected from their double-lined spectra or from IR excess. But EHB stars with WD or M-dwarf companions show no such features. When the orbital periods are sufficiently short, these systems can easily be revealed from their RV variations. Using the RV method, Maxted et al. (2001) targeted 36 EHB stars and found 21 binaries, all with periods less than 30 days. This gave a fraction of short period binaries of $60 \pm 8\%$. Other surveys have found smaller fractions, but they have not constrained the sample to focus strictly on the EHB. From high-resolution VLT spectra of 76 sdB stars from the SPY survey Lisker et al. (2005) found that 24 showed the signature of an FGK companion,

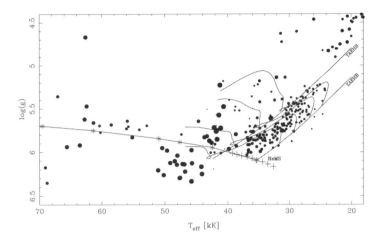

Figure 1: The EHB in the $T_{\rm eff}$–$\log g$ plane as observed by the Bok-Green survey (Green et al. 2008). The symbols mark observed stars with the size indicating the helium abundance. The theoretical zero age HeMS is shown for a wide range of masses. Models from Paczyński (1971) with masses of 0.5, 0.7, 0.85, 1, 1.5 and 2 $M_\odot$ are marked with ∗ symbols (starting from low $T_{\rm eff}$). More recent models from Kawaler & Hostler (2005) are shown for $M_* = 0.41, 0.43, ...0.57$ $M_\odot$ marked with + symbols, and the zero age and terminal age EHB for the 0.47 $M_\odot$ core models are also drawn. For the latter, four evolutionary tracks with different envelope thicknesses $\log M_e/M_* = -3.5, -3, -2.5, -2$ are drawn (starting from high $\log g$).

none of which show any RV variability. Napiwotzki et al. (2004) reported that of 46 sdB stars in the same sample, 18 (39%) were RV variable. Clearly, the binary fraction in EHB stars is much higher than for normal stars, but an accurate number has yet to be established.

Most recently, Green et al. (2008) have presented a uniform high signal-to-noise low-resolution survey of a large sample including most known hot subdwarf stars brighter than $V = 14.2$, using the university of Arizona 2.3 m Bok telescope (hereafter referred to as the Bok-Green or BG survey). From this large sample the clearest picture of the EHB to date emerges (Figs. 1 and 2). Most stars in the diagram are clearly well bound by EHB models for a narrow mass distribution. Most of the remaining stars are consistent with post-EHB models, but could also fit core helium burning objects with higher than canonical masses. The most helium rich objects, however, appear to form their own sequence, which cannot be explained by canonical EHB models. The tail of the regular horizontal branch reaches into the diagram from the upper right, but is separated from the EHB by a substantial gap, as first noted by Newell (1973). Although the details of this survey are still under analysis, several new features have been noted. The sequence of He-rich objects around 40 kK is not compatible with current evolutionary scenarios, since post-EHB and post-RGB objects pass too rapidly through this region of the $T_{\rm eff}$–$\log g$ plane to produce the observed clustering, but the *late hot flasher* scenario (discussed in the next section) holds some promise.

The large group of stars below the helium main sequence (HeMS) is more problematic as no type of star should be able to stay in this position in the $T_{\rm eff}$–$\log g$ plane more than briefly, and no clustering should occur. The feature was also noted by Stroeer et al. (2007), but it cannot be ruled out that this is an artifact of the models, as many of these stars appear to have significant amounts of CNO processed material in their atmospheres, and the NLTE models used do not account for this. Another remarkable feature appears when looking at the distribution of short period binaries in the $l_{\rm eff}$–$\log g$ plane (Fig. 2). In particular, the incidence of such binaries appears to be much smaller at the hot, high gravity end of the

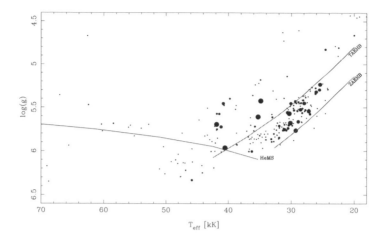

Figure 2: The same as Fig. 1, but with the symbol size indicating the dispersion in radial velocity. The EHB stars with the highest velocity variations appear to be concentrated at lower gravities on the EHB. sdB+FGK stars are not included here, due to difficulties in reliably disentangling such composite spectra.

EHB strip than among the cooler subdwarfs. This can be understood in terms of the relative efficiency of the common envelope ejection channel (CE, see next section) that produces short period binaries versus the other channels producing long period systems or single stars. It would appear that the CE channel is significantly less effective in removing the envelope than the other channels.

Formation and evolution

As binary interactions are a key for understanding the formation of EHB stars, I will attempt a short introduction here. Mass transfer during close binary evolution is well understood, although there are still some unknown factors. While the full picture is exceedingly complex, and would take far too much time to describe here, I will try to outline some of the most important possibilities.

Losing the Envelope

There are two fundamental evolutionary paths, and which one a system enters depends only on the mass ratio of the system. If the expanding mass donor is more massive than the accretor, the orbit will shrink catastrophically and the system enters a common envelope (CE) phase. As the orbit shrinks further due to friction, orbital energy is deposited in the envelope, spinning it up. When sufficient energy is deposited the envelope is ejected. Stars with an initial main sequence (MS) mass below about $1.8\,M_\odot$ can ignite helium in a core flash before the tip of the RGB, and if the envelope is ejected at the right time the result is an EHB star with a mass close to the flash mass of $\sim 0.47\,M_\odot$, and a very close low-mass companion. If the envelope is ejected before the core reaches the required mass, the core never ignites helium and the star will not settle on the EHB, but continues to contract and ends up as a helium core white dwarf. The close sdB+WD binary HD 188112 (Heber et al. 2003) is a particular point case, as the Hipparcos parallax together with the observed spectroscopic surface gravity clearly constrain the mass of the subdwarf component to be below the core helium burning limit. If the MS mass is higher than about $2.0\,M_\odot$ the star will ignite helium non-degeneratively well before the core reaches the mass required for the helium flash. If the envelope is subsequently ejected on the tip of the RGB the outcome would be an EHB star which could have a mass as low as $0.33\,M_\odot$.

On the other hand, if the companion is more massive than the red giant donor filling its Roche lobe, the orbit expands and no CE is formed. In this stable Roche lobe overflow (RLOF) scenario the orbital period can end up as long as 2000 days. As with CE ejection, if the red giant starts out with a mass below the mass required for the helium flash to occur, the mass transfer must happen close to the tip of the RGB, and the core of the giant becomes an EHB star with a mass close to $0.47\,M_\odot$. If it is too massive, non-degenerate helium burning starts earlier, and the result is an sdB star with a mass between 0.33 and $1.1\,M_\odot$ (Han et al. 2000).

The Problematic Singles

A significant number of sdB stars are definitely single stars, and their formation is the most problematic and controversial. While post-CE systems leave behind a close binary that is easily detectable from the radial velocity (RV) variations, post-RLOF binaries have such long periods that it requires very long term efforts with high precision spectroscopy to detect them. Up to now there are no detections of any such orbits. However, long term asteroseismic monitoring can detect orbitally induced variations in the pulsation period with much higher precision than can be done from spectroscopy. The clearest case yet is V391 Peg where the modulation of the pulsations are consistent with a planet with a mass ($M \sin i$) of $3.2\,M_{Jupiter}$ in an orbit with period of 1170 days (Silvotti et al. 2007). While the planet might have entered the outer layers of the envelope of the red giant before the envelope was lost, the current orbit is too wide for it to have been responsible for the actual envelope ejection.

Several scenarios have been proposed that may produce single EHB stars. It is well known that RGB stars lose significant mass in the form of a stellar wind as they expand and their surface gravity becomes extremely low. D'Cruz et al. (1996) computed evolutionary models of RGB stars with mass loss parameterized by the Reimers efficiency η_R. They found that the observed distributions of HB and EHB stars can be explained *"so long as nature provides a broad enough distribution in η_R"*. However, the actual physics behind the large variation in the mass loss efficiency remains unexplained. Another possibility is the merger of two He-core WD stars, first proposed by Webbink (1984). Saio & Jeffery (2000) have shown that models for such a merger can successfully predict the behaviour of the pulsating helium star V652 Her, demonstrating the feasibility of the merger scenario. Such extreme helium stars will evolve to become hot helium rich subdwarfs located close to the HeMS. However, a remaining problem with merger models is that they invariably leave behind rapidly rotating objects. So far, no single hot subdwarf has demonstrated more than moderate rotational velocities from high resolution spectroscopy.

Another possibility is that CE ejection can be triggered by a giant planet that evaporates in the process (Soker 1998). Nelemans & Tauris (1998) demonstrated in the context of white dwarf evolution that there are clear domains in initial orbital period versus planetary mass, where the planet ejects the envelope and is disrupted as it fills its own Roche-lobe after the spiral in. The final rotation period of the remaining helium core is not affected by this process, as the planet transfers almost all of its angular momentum to the envelope before its ejection. A final possibility for the formation of single hot subdwarfs, also noted by Nelemans & Tauris (1998) in the context of formation of undermassive white dwarfs, is a variation of the RLOF mechanism. If the envelope is transferred onto an accretor that is already a massive white dwarf, an asymmetric accretion induced collapse may produce a high velocity neutron star which escapes the system. If the companion is sufficiently massive to explode as a supernova, Marietta et al. (2000) have computed that the explosion itself impacts 1000 times more energy on the envelope than its binding energy, easily stripping the giant to the core. If the core is massive enough for helium burning, and the SN explosion sufficiently asymmetric, the remnant of the mass donor would end up as a single EHB star. In both cases the disruption of such a binary system would leave the EHB star with an unusual galactic orbit, which should be observable at least in a sufficiently large sample.

To Flash or Not to Flash

If an RGB star loses its envelope before the core has reached the mass required for the helium flash, the core will contract and heat up, before cooling as a helium core WD. The tracks calculated for such flashless post-RGB evolution covers a wide span in temperatures, with models around $0.2\,M_\odot$ passing through the cool end of the EHB, and the remnants with masses close to the helium flash mass reaching temperatures up to 100 kK (Driebe et al. 1999).

A borderline case exists when the mass is just on the limit for the helium core flash to occur. Then the flash can happen after the RGB stage, while the core is either on its way to or on the actual WD cooling curve. Such models are known as hot flashers, and the eventual outcome depends on the exact stage at which the helium flash occurs. If ignition occurs before the turning point on the WD cooling track (early hot flashers), the remaining H-burning shell produces a sufficient entropy barrier to prevent the convection zone produced by the helium flash to reach the surface (Iben 1976). But if helium ignites on the actual WD cooling curve, any remaining shell H-burning is too weak to prevent the convection zone reaching the envelope. The envelope hydrogen is then mixed into the core and quickly burnt (Sweigart 1997), and CNO processed material is transported to the outer layers in a flash mixing process. Such *late hot flashers* are predicted to end up with an atmosphere almost totally devoid of hydrogen and with observable CNO lines in their spectra.

Recently, Miller Bertolami et al. (2008) have performed extensive simulations of late hot flashers in order to determine how well models can reproduce observations. They predict that the core flash cycles should take place in the region above the HeMS where the strip of helium rich subdwarfs are concentrated (Fig. 1). However, the core flash phase lasts less than 2 Myr, after which the stars settle close to the HeMS, for a regular EHB lifetime of at least 66 Myr. Observations do not support such a concentration of objects at this location in the T_{eff}–$\log g$ plane. To resolve this they propose that some remaining hydrogen could have survived the mixing and should slowly diffuse to the surface, effectively pulling the star up toward the cooler region of the EHB.

Another very recent development was presented by Politano et al. (2008). They have extended the classic common envelope ejection mechanism to include the case when a low-mass MS star or brown dwarf merges with the helium core, in order to produce single EHB stars. However, as with the helium white dwarf merger scenario, the problem remains that the products end up spinning close to break-up velocity. Since a rapidly rotating subpopulation of sdB stars has yet to be found, this channel is only of marginal interest, unless a way to eject the envelope without spinning up the core can be found.

Asteroseismology

The first evidence of rapid pulsations in EHB stars were reported by Kilkenny et al. (1997), after their detection of multiperiodic pulsations in V361 Hya. The V361 Hya stars span the hot end of the EHB strip from about 28 to 34 kK, and pulsate in p modes of low l orders and with photometric amplitudes up to 6 %. The pulsation periods range between 100 and 400 s, and 40 such stars are known in the literature to date (Oreiro et al. 2009). One of these, V338 Ser, has periods reaching almost 10 minutes, but stands out as it sits well above the EHB (Fig. 3), being most likely in a post-EHB stage of evolution.

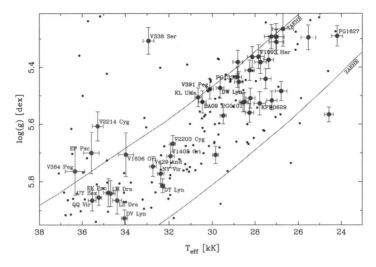

Figure 3: Section of the T_{eff}–log g plane where the EHB stars are located. Pulsators with temperatures and gravities in the BG survey are marked with big symbols and error bars. Small symbols without error bars are stars not observed to pulsate. In the online version the colours indicate short period pulsators with (*green*) and without (*red*) published asteroseismic solution, long period pulsators (*magenta*) and hybrid pulsators (*blue core*).

It took several years from the announcement of the first sdB pulsator until it was realized that the same group of stars are subject to longer period pulsations as well. Green et al. (2003) published the discovery of pulsations in V1093 Her, with periods between one half and two hours, and reported that as many as 75% of sdB stars cooler than 30 kK display some level of pulsations at these periods. The V1093 Her stars span the EHB from the coolest sdBs at around 24 kK up to the domain of the V361 Hya stars (Fig. 3). The pulsations were identified with high radial order g modes by Green et al. (2003), and their amplitudes are very low, typically 0.2 % or less. Although such low level pulsations are common in V1093 Her stars, and have been detected in at least 30, only a few have been studied in detail due to the long time-base and high precision required to detect and resolve their modes.

Even more recently, Schuh et al. (2006) realized that a known V361 Hya star, DW Lyn, was actually displaying the g modes of a V1093 Her star simultaneously with p modes of a V361 Hya star. As noted by the authors, the four V361 Hya stars DW Lyn, V391 Peg and Balloon 090100001 (BA09 in Fig. 3 and hereafter), and KL UMa, form a compact group closer to the domain of the V1093 Her stars than the remaining V361 Hya stars. Hybrid DW Lyn type pulsations have now been detected in both V391 Peg and BA09 as well, but appear to be absent in KL UMa. Intriguingly, KL UMa also stands out as the only binary of the quartet (O'Toole et al. 2004).

A fourth type of pulsations was noted in the He-sdB star LS IV–14°116 by Ahmad & Jeffery (2005). They detected pulsations with amplitudes at the 1% level and periods around 15 minutes. The atmospheric parameters reported by the authors, $T_{eff} = 32.5$ kK, log $g = 5.4$ dex places the star just above the EHB strip in the T_{eff}–log g diagram, well surrounded by V361 Hya stars. With a supersolar helium abundance, log(He/H) = -0.6, this star represents a different evolutionary state than the regular EHB pulsators. Up to now this star remains unique, but since stars with the atmospheric properties of LS IV–14°116 are extremely rare it is too early to tell whether pulsations in stars with similar atmospheric parameters are common or rare.

A fifth and final class of pulsations in hot subdwarf stars was discovered by Woudt et al. (2006) in the hot sdO binary J17006+0748, but this will not be discussed here as this star is very far from the EHB region.

Driving the Beat

Pulsations in sdB stars were predicted to occur by Charpinet et al. (1996) at about the same time as the first pulsators were discovered by Kilkenny et al. (1997). The driving mechanism is due to an opacity bump caused by iron group elements (Charpinet et al. 1997). This mechanism is inefficient at solar metallicity, but gravitational settling and radiative levitation can work together to locally enhance metals in a driving zone in the envelope. This κ mechanism has been successfully invoked to explain both the p-mode pulsations in V361 Hya stars and the g-mode pulsations in V1093 Her stars (Fontaine et al. 2003).

While the first models by Fontaine et al. (2003) could produce unstable modes in the coolest sdB stars, there appeared to be a gap between one island of instability on the cool end of the EHB and one at the hot end. The pulsators at the hot end of the g-mode instability region remained problematic, and particularly so the hybrid DW Lyn type pulsators. Some relief to this problem was recently provided by Jeffery & Saio (2007), with the application of improved opacity values from the OP project as well as the explicit consideration of nickel in addition to iron. The new models are sufficient to bridge the gap between the hot and cool EHB pulsators, and could also predict an island of instability in the sdO domain, close to the observed location of J17006+0748. However, these most recent calculations do not yet give a perfect description of the observed picture. More unstable modes are still found in models at the hot end of the EHB than on the cool end, while observations indicate that pulsations are more common in cool sdB stars. In fact, the problem is now more to explain why most EHB stars on the hot end of the strip do not pulsate, than why they do. Jeffery & Saio (2007) speculated that the iron group element enhancements, which build up due to a diffusion process over rather long time scales, may be disrupted by the the atmospheric motions as pulsations build up to some level. They note that since p modes mostly involve vertical motion, while g modes are dominated by horizontal motion, it is possible that p modes are more effective at redistributing the iron group elements out of the driving zone.

Levitation does the Trick

Detailed models of the internal structure of the EHB stars are critical for improving our understanding of the asteroseismic properties. The simplest models with uniform metallicities are not able to drive pulsations in these stars at all. Only with the inclusion of an iron opacity bump was it possible to find unstable modes (Charpinet et al. 1996). However, the periods predicted by these so-called second generation models have usually not been matched with observed periods to better than about one percent, while the observed periods have a precision that are an order of magnitude better. Efforts to improve the atmospheric models to include more of the various effects that can have significant impact on the pulsation spectrum are ongoing.

Important progress was reported by Fontaine et al. (2006), who clearly demonstrated the importance of properly including time-dependent diffusion calculations in order to predict pulsation frequencies and mode stability. Starting from a uniform distribution and solar iron abundance, they demonstrated that it takes several hundred thousand years for radiative acceleration and gravitational settling to produce sufficient iron in the driving region to create unstable modes. After about 1 Myr there are no more changes with respect to which modes are driven and not, but the pulsation frequencies may still shift as the iron opacity bump builds up further. After about 10 Myr iron reaches equilibrium in these models, and no

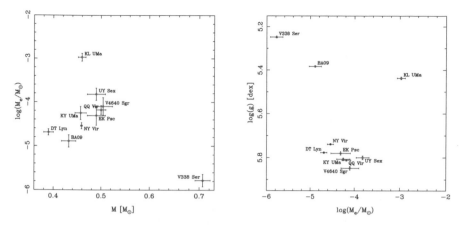

Figure 4: Published asteroseismic solutions for ten V361 Hya stars. The left hand panel shows total mass plotted versus envelope mass, and the right hand panel shows envelope mass versus surface gravity. Note that only the optimal solution is shown even if the papers discuss several possible ones.

further changes are seen. Since the time to onset of pulsations is just $1/1000$ of the typical EHB lifetime, this does not help resolve the issue as to why only a fraction of the EHB stars pulsate.

Second generation models based on a pure hydrogen atmosphere on top of a simple 'hard ball' core approximation, to which an explicit iron abundance profile is added, have been used to derive sensible asteroseismic quantities for a number of V361 Hya stars (e.g. Charpinet et al. 2007). The adopted 'forward' method basically consists of constructing a large grid of models in the four dimensional parameter space spanned by the fundamental model parameters, effective temperature T_{eff}, surface gravity $\log g$, total mass M, and envelope mass fraction $q(H)$. A minimization procedure is then invoked to find the model that best matches the observed periods.

To date, ten asteroseismic solutions computed with this method have been published. They were summarized in Randall et al. (2007) for the first seven, and in Fig. 4 the new solutions by van Grootel et al. (2008), van Spaandonk et al. (2008) and Charpinet et al. (2008) (discussed below) have been included. A feature of the asteroseismic modelling is that T_{eff} is rather poorly constrained, and a better value can usually be provided from spectroscopy. The surface gravity, total mass, and envelope mass fraction, however, have very small associated errors in the asteroseismic solutions, so we plot only these in Fig. 4. The distribution of masses is not as concentrated around $0.47\,M_\odot$ as most canonical evolutionary models have presumed, but all points are well within the permitted ranges for synthetic populations considered by Han et al. (2002, 2003). Except for two outliers, all the stars appear to form a trend with envelope mass, M_e, increasing with total mass, M. Although this feature has not been accounted for by evolutionary calculations, it could occur as a natural consequence of a higher core mass requiring more energy to remove the envelope. More disturbing is the lack of any clear trend in envelope mass versus surface gravity, as is clearly demanded for canonical EHB models. The scatter in the high gravity objects is easily explained by their spread in mass, and KL UMa fits well with the expected $M_e/\log g$ trend. But the unusually low envelope masses for BA09 and V338 Ser are hard to explain, and may indicate that the adopted models are too simplified to represent the seismic properties for these cases.

Ballooney

The exceptional amplitude of the dominant period in BA09 has hinted towards a radial nature, and this was finally confirmed by Baran et al. (2008) by combining evidence from multicolour photometry and the radial velocity amplitude measured by Telting & Østensen (2006).

Van Grootel et al. (2008) have successfully applied the forward method to BA09, demonstrating some peculiarities in the model predictions. Their optimal solution for the main mode, when using no constraints, is $l = 2$, which is not reconcilable with the spectroscopic data. However, by imposing mode constraints from multicolour photometry they do find asteroseismic solutions that agree with all observational data. Curiously, the physical parameters for the constrained and unconstrained fits are almost identical, even if the mode identification changes for half the modes considered. The authors conclude *"Our primary result is that the asteroseismic solution stands very robust, whether or not external constraints on the values of the degree l are used."* This peculiarity arises from the high mode density and the way the modes are distributed in period space. But the deeper cause of the problem is the low precision with which the second generation models predict pulsation periods. Models with a more detailed internal structure are therefore urgently needed in order to resolve this problem. It is a concern that the envelope mass fraction van Grootel et al. (2008) find ($\log M_e/M_* \simeq -4.9$, regardless of which modes are which) is several hundred times lower than what any EHB model would predict for such a low-mass star at this position in the T_{eff}–$\log g$ diagram. With such a thin envelope all models put the star close to the HeMS for a core helium burning star. The authors' suggestion that the star is in a post-EHB stage of evolution is beyond canonical theory, as only models with substantial hydrogen envelopes evolve to lower gravities before moving off to the sdO domain as the core starts to contract (see Fig. 1).

The envelope mass discrepancy is even more severe in the asteroseismic results for V338 Ser (van Spaandonk et al. 2008), whose best fitting model (number 4 of 5 presented) has an envelope mass fraction $\log M_e/M_* = -5.78$! While the authors seem to prefer an even more extreme value of −6.22 for a model with a slightly poorer fit, in order to obtain a mass of $0.561\,M_\odot$ rather than the unusually high mass of $0.707\,M_\odot$, the high-mass solution might be the most interesting. For if the exceptionally high mass is real, evolutionary calculations would place V338 Ser in a core helium burning stage, not in a post-EHB stage as would be the case if its mass was around $0.5\,M_\odot$. But as for BA09, evolutionary models demand an envelope mass fraction more than 1 000 times higher than found by van Spaandonk et al. (2008), in order to find V338 Ser at the observed $\log g$.

With a recent update of the forward modelling code, Charpinet et al. (2008) have produced a very convincing model for the eclipsing binary system NY Vir. This star has been particularly challenging since it is rapidly rotating, due to being in a tidally locked orbit with the close M-dwarf companion. The rotational splitting of modes with different m produces a particularly rich pulsation spectrum. Charpinet et al. (2008) use asteroseismology to discriminate between three solutions from the binary orbit published by Vučković et al. (2007), and find that the intermediate model with a mass of $0.47\,M_\odot$ is clearly favoured. This solution is also the only one consistent with the $\log g$ from the BG survey (Fig. 3).

Most recently, Telting et al. (2008) have presented the first study of line-profile variations in these stars based on high-resolution spectroscopy. Line profile variations of metal lines is a technique to directly determine the spherical harmonic order numbers l, m of a pulsation mode, which is well established for various MS pulsators. To invoke this technique on the faint EHB stars requires substantial investments in terms of telescope time, which has prohibited its use up to now. With the preliminary results on the high amplitude pulsator BA09, Telting et al. (2008) clearly demonstrate that the l of the main mode must be either 0 or 1. Again, the observational accuracy has advanced ahead of the theoretical models, as the standard modelling of such line profile variations are insufficient to properly account for the complex effects on the line profiles in the high temperature and gravity domain of the EHB stars.

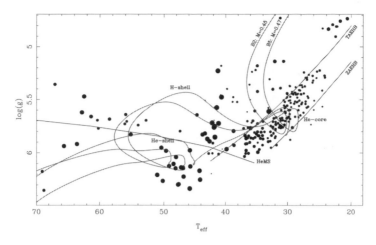

Figure 5: Same as Fig. 1, but with the evolutionary tracks for flashless post-RGB evolution from Hu et al. (2008) overplotted. The tracks evolve rapidly from the top of the plot to the ZAEHB as the envelope settles down on the helium burning core, and the stars spends most of their time (200 Myr) in the upward part of the loop at constant $T_{\rm eff}$. After core helium exhaustion the star rapidly heats up, but then turns and moves back again for a several Myr long shell helium burning phase below the HeMS.

A final advancement that demonstrates with particular clarity the direction in which asteroseismology of EHB stars needs to move in order to progress, is the work of Hu et al. (2008). The authors have taken real evolutionary models that have been evolved from the ZAMS, through flashless helium ignition on the RGB, to the EHB by peeling of the envelope, and proceeded to compute the pulsation properties of these stars for a number of different configurations. By comparing these models with the classical post-flash models with similar surface parameters, they clearly show that the differences in internal structure from the two evolutionary paths produce significant differences in both the predicted pulsation periods, and with respect to which modes are excited or damped! Two of the flashless models of Hu et al. (2008) are plotted in Fig. 5 together with a classical post-flash EHB model. From an evolutionary population synthesis point of view it is interesting to note that the post-EHB tracks are substantially different. After the core has exhausted its helium and the star moves off the TAEHB the star briefly burns the remaining envelope hydrogen, before contracting and cooling down during a helium shell burning phase, which lasts much longer than in canonical models (up to 20 Myrs). This is long enough to produce a slight clustering of objects below the HeMS at temperatures between 45 and 50 kK, just where a substantial cluster of He-sdO stars are observed.

Conclusions

Asteroseismology of EHB stars is a field in rapid progress, with exceptional challenges due to their complex formation paths. Much progress has been made on evolutionary models, but much remains to be done, particularly with respect to the formation of single EHB stars. Better models are also needed to reproduce the effects of the helium flash stage on the envelope composition, in order to reproduce the internal structure of EHB stars at the accuracy achieved by observational asteroseismology. Only when these advances are in place can asteroseismology reliably test the different formation scenarios.

The low amplitudes and long periods make it very difficult to establish detailed pulsation spectra for V1093 Her stars, and even when it can be done the high mode density makes it difficult to assign modes to the observed frequencies. But g modes are particularly interesting because they probe deep into the stellar interior. This is a significant challenge for the future due to the long time-base required to reliably determine the longer pulsation periods in these stars. The upcoming Kepler satellite mission (Christensen-Dalsgaard et al. 2007) provides an excellent opportunity if pulsators can be found within its field of view.

Acknowledgments. Special thanks to E.M. Green for kindly providing the detailed results from her 2008 article, which made it possible to reproduce the figures from that article together with evolutionary tracks. Without these data, Figs. 1, 2, 3, and 5 would have been a lot less informative. The author is supported by the Research Council of the University of Leuven under grant GOA/2008/04 and by the EU FP6 Coordination Action HELAS.

References

Ahmad, A., & Jeffery, C. S. 2005, A&A, 437, L51

Baran, A., Pigulski, A., & O'Toole, S. J. 2008, MNRAS, 385, 255

Charpinet, S., Fontaine, G., Brassard, P., et al. 1997, ApJ, 483, L123

Charpinet, S., Fontaine, G., Brassard, P., & Dorman, B. 1996, ApJ, 471, L103

Charpinet, S., Fontaine, G., Brassard, P., et al. 2007, CoAst, 150, 241

Charpinet, S., van Grootel, V., Reese, D., et al. 2008, A&A, 489, 377

Christensen-Dalsgaard, J., Arentoft, T., Brown, T. M., et al. 2007, CoAst, 150, 350

D'Cruz, N. L., Dorman, B., Rood, R. T., & O'Connell, R. W. 1996, ApJ, 466, 359

Driebe, T., Blöcker, T., Schönberner, D., & Herwig, F. 1999, A&A, 350, 89

Edelmann, H., Heber, U., Hagen, H.-J., Lemke, M., et al. 2003 A&A, 400, 939

Fontaine, G., Brassard, P., Charpinet, S., et al. 2003, ApJ, 597, 518

Fontaine, G., Brassard, P., Charpinet, S., & Chayer, P. 2006, Mem. S.A.It., 77, 49

Green, E. M., Fontaine, G., Reed, M. D., et al. 2003, ApJ, 583, L31

Green, E. M., Fontaine, G., Hyde, E. A., et al. 2008, ASP Conf. Ser., 392, 75

Green, R. F., Schmidt, M., & Liebert, J. 1986, ApJSS, 61, 305

Greenstein, J. L., & Sargent, A. I. 1974, ApJSS, 28, 157

van Grootel, V., Charpinet, S., Fontaine, G., et al. 2008, A&A, 488, 685

Han, Z., Tout, C. A., Eggleton, P. P. 2000, MNRAS, 319, 215

Han, Z., Podsiadlowski, Ph., Maxted, P. F. L, et al. 2002, MNRAS, 336, 449

Han, Z., Podsiadlowski, Ph., Maxted, P. F. L, & Marsh, T. R. 2003, MNRAS, 341, 669

Heber, U. 1986, A&A, 155, 33

Heber, U., Edelmann, H., Lisker, T., Napiwotzki, R. 2003, A&A, 411, L477

Heber, U. 2009, Annual Review of Astronomy and Astrophysics, 47, submitted

Hu, H., Dupret, M.-A., Aerts, C., et al., 2008, A&A 490, 243

Humason, M. L., & Zwicky, F. 1947, ApJ, 117, 313

Iben Jr, I. 1976, ApJ, 208, 165

Jeffery, C. S., & Saio, H. 2007, MNRAS, 378, 379

Kawaler, S. D., & Hostler, S. R. 2005, ApJ, 621, 432

Kilkenny, D., Koen, C., O'Donoghue, D., & Stobie, R.S. 1997, MNRAS, 285, 640

Lisker, T., Heber, U., Napiwotzki, R., et al. 2005, A&A, 430, 223

Marietta, E., Burrows, A., Fryxell, B., 2000, ApJS 128, 615

Maxted, P. F. L, Heber, U., Marsh, T. R., & North, R. C. 2001, MNRAS, 326, 1391

Miller Bertolami, M. M., Althaus, L. G., Unglaub, K., & Weiss, A. 2008, A&A, 491, 253

Moehler, S., Richtler, T., de Boer, K. S., et al. 1990, A&ASS, 86, 53

Napiwotzki, R., Karl, C. A., Lisker, T., et al. 2004, Ap&SS, 291, 321

Nelemans, G., & Tauris, T. M. 1998, A&A, 335, 85

Newell, E. B. 1973, ApJSS, 26, 37

Oreiro, R., Østensen, R. H., Green, E. M., & Geier, S. 2009, A&A, in press

Østensen, R. H. 2006, Baltic Astronomy, 15, 85

O'Toole, S. J., Heber, U., & Benjamin, R. A. 2004, A&A, 422, 1053

Paczyński, B. 1971, Acta Astronomica, 21, 1

Podsiadlowski, Ph., Han, Z., Lynas-Gray, A. E., & Brown, D. 2008, ASP Conf. Ser., 392, 15

Politano, M., Taam, R. E., van der Sluys, M., & Willems, B. 2008, ApJ, 687, L99

Saio, H., & Jeffery, C. S. 2000, MNRAS, 313, 671

Schuh, S., Huber, J., Dreizler, S., Heber, U., et al. 2006, A&A, 445, L31

Silvotti, R., Schuh, S., Janulis, R., Solheim, J.-E., et al. 2007, Nature, 449, 189

Soker, N. 1998, AJ, 116, 1308

van Spaandonk, L., Fontaine, G., Brassard, P., & Aerts, C. 2008, ASP Conf. Ser., 392, 387

Stoughton, C., Lupton, R.H., Bernardi, M., et al. 2002, AJ, 123, 485

Stroeer, A., Heber, U., Lisker, T., et al. 2007, A&A, 462, 269

Sweigart, A. V. 1997, in "The Third Conference on Faint Blue Stars", eds. A. G. D. Philip, J. Liebert, R. Saffer, and D. S. Hayes, L. Davis Press, p. 3, arXiv:astro-ph/9708164

Telting, J. H., & Østensen, R. H. 2006, A&A, 450, 1149

Telting, J. H., Geier, S., Østensen, R. H., Heber, U., et al. 2008, A&A, 492, 815

Vučković, M., Aerts, C., Østensen, R. H., et al. 2007, A&A, 471, 605

Webbink, R. F. 1984, ApJ, 277, 355

Woudt, P. A., Kilkenny, D., Zietsman, E., et al. 2006, MNRAS, 371, 1497

Comm. in Asteroseismology,
Vol. 159, 2009, JENAM 2008 Symposium № 4: Asteroseismology and Stellar Evolution
S. Schuh & G. Handler

A search for Extreme Horizontal Branch pulsators in ω Cen

S. K. Randall,[1] A. Calamida,[1] and G. Bono[2,3]

[1] ESO, Karl-Schwarzschild-Str. 2, 85748 Garching bei München, Germany
[2] Istituto Nazionale de Astrofisica, Osservatorio Astronomico di Roma,
Via Frascati 33, 00040 Monte Porzio Catone, Italy
[3] Università di Roma "Tor Vergata", Department of Physics,
Via della Ricerca Scientifica 1, 00133, Rome, Italy

Abstract

We report the discovery of a new pulsating EHB star from a search for rapidly pulsating Extreme Horizontal Branch in the globular cluster ω Cen.

Individual Objects: ω Cen

We report on the outcome of a search for rapidly pulsating Extreme Horizontal Branch (EHB) stars in ω Cen on the basis of 2 hours of SUSI2 rapid time-series photometry gathered at the 3.5-m NTT on La Silla, Chile. The field observed covers 5.5'×5.5' in the southeastern quadrant of ω Cen, which was selected over more typical globular clusters for its well-populated EHB, as well as its relative proximity and low reddening. We used a U filter in order to minimize field crowding, and chose 3x3 binning to reduce the overhead time to 16 s, which combined with the exposure time of 20 s resulted in a cycle time of 36 s, low enough to detect the rapid oscillations expected in EC 14026 type stars, the main targets of our variability search.

These objects make up a small fraction ($\sim$ 5%) of subdwarf B (sdB) stars, which are evolved, core-helium burning objects located on the EHB of the H-R diagram. While they are thought to be the progeny of stars that suffered significant mass loss near the tip of the Red Giant Branch, the details of their formation remain unclear. It is hoped that eventually, the asteroseismic interpretation of the pulsators among them will enable a characterization of the mass and hydrogen-shell thickness distribution of the sdB population and thus help discriminate between different proposed evolutionary scenarios. First efforts in this direction appear promising, full asteroseismic analyses having so far been carried out for 12 out of at least 35 known EC 14026 stars (see Fontaine et al. 2008 for a recent review). However, until now, all known sdB pulsators belonged to the field population, despite several searches for variability among EHB stars in selected globular clusters.

The field we monitored with SUSI2 completely overlaps with available UBVI-band photometry of ω Cen gathered with WFI on the 2.2-m ESO/MPI telescope (Castellani et al. 2007). Performing simultaneous PSF-fitting photometry on the 192 SUSI2 frames obtained led to the detection of $\sim$ 20,000 stars, of which we were able to select potential EC 14026 star candidates from the WFI catalogue in terms of brightness ($16 \leq U \leq 18.5$), colour ($-2 \leq (U - V) \leq -0.8$), photometric accuracy, sharpness and separation index. For the 52 EHB stars thus identified we computed the airmass and seeing corrected light curves with respect to the mean SUSI2 u-band magnitude. Note that the latter was not calibrated and does therefore not constitute a standard magnitude. We then computed the Fourier transform

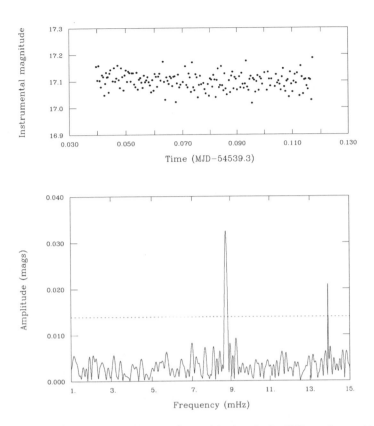

Figure 1: Light curve (top) and Fourier transform (bottom) for the pulsating EHB star discovered in ω Cen. Note that the relative u magnitude indicated for the light curve is offset by around +0.25 magnitudes with respect to the absolute U magnitude from the WFI catalogue. The dominant pulsation has a period of 114 s, while the strange-looking secondary peak corresponds to the Nyquist frequency. The horizontal dashed line indicates the 4σ detection threshold.

(FT) for each light curve in the 1−15 mHz range, appropriate for detecting the pulsations expected for the EC 14026 pulsators.

Of the 52 selected candidates, one shows a credible peak in the FT above the imposed detection threshold of 4σ. Its light curve and Fourier transform are displayed in Fig. 1, where both the 114-s (8.75 mHz) peak thought to indicate stellar oscillation as well as an observational artefact at 72 s (13.9 mHz) corresponding to the Nyquist frequency are visible. The latter is encountered in most of the targets monitored, while the former is unique to the star displayed. This strengthens the case for the discovery of real variability in the star rather than an instrumental or observational signature.

The period detected for the variable star ties in well with the typical 100−200 s pulsations observed for EC 14026 stars. Moreover, a comparison of the observed optical colours with Horizontal Branch models appropriate for ω Cen indicates an effective temperature of 31,500 ± 6,300 K for this object, placing it well within the 29,000−36,000 K instability strip for rapidly oscillating subdwarf B stars. Therefore, we are quite confident that we have discovered the first EC 14026 star in a globular cluster.

Further details on the work presented here can be found in Randall et al. 2009.

Acknowledgments. S.K.R. would like to thank the ESO La Silla staff, in particular SUSI2 instrument scientist Alessandro Ederoclite. Sadly, the observations reported here were among the last ever obtained with SUSI2 since the instrument has since been decommissioned.

References

Castellani, V., Calamida, A., Bono, G., et al. 2007, ApJ, 663, 1021

Fontaine, G., Brassard, P., Charpinet, S., et al. 2008, ASP Conf. Ser., 392, 231

Randall, S. K., Calamida, A., & Bono, G. 2009, A&A, 494, 1053

Comm. in Asteroseismology,
Vol. 159, 2009, JENAM 2008 Symposium № 4: Asteroseismology and Stellar Evolution
S. Schuh & G. Handler

Time-resolved spectroscopy of the planet-hosting sdB pulsator V391 Pegasi

S. Schuh,[1] R. Kruspe,[1] R. Lutz,[1] and R. Silvotti[2]

[1] Institut für Astrophysik, Universität Göttingen, Friedrich-Hund-Platz 1, 37077 Göttingen, Germany
[2] INAF - Osservatorio Astronomico di Capodimonte, via Moiariello 16, 80131 Napoli, Italy

Abstract

The subdwarf B (sdB) star V391 Peg oscillates in short-period p modes and long-period g modes, making it one of the three known hybrids among sdBs. As a by-product of the effort to measure secular period changes in the p modes due to evolutionary effects on a time scale of almost a decade, the O−C diagram has revealed an additional sinusoidal component attributed to a periodic shift in the light travel time caused by a planetary-mass companion around the sdB star in a 3.2 yr orbit. In order to derive the mass of the companion object, it is necessary to determine the orbital inclination. One promising possibility to do this is to use the stellar inclination as a primer for the orbital orientation. The stellar inclination can refer to the rotational or the pulsational axis, which are assumed to be aligned, and can in turn then be derived by combining measurements of v_{rot} and $v_{rot} \sin i$.

The former is in principle accessible through rotational splitting in the photometric frequency spectrum (which has however not been found for V391 Peg yet), while the projected rotational velocity can be measured from the rotational broadening of spectral lines. The latter must be deconvolved from the additional pulsational broadening caused by the surface radial velocity variation in high S/N phase averaged spectra. This work gives limits on pulsational radial velocities from a series of phase resolved spectra.

Phase averaged and phase resolved high resolution echelle spectra were obtained in May and September 2007 with the 9m-class Hobby-Eberly Telescope (HET), and one phase averaged spectrum in May 2008 with the 10m-Keck 1 telescope[1].

Individual Objects: V391 Pegasi

The hybrid pulsating sdB star V391 Pegasi and its planetary companion

Subdwarf B stars (sdBs) are subluminous, evolved stars on the extreme horizontal branch (EHB). They have a He burning core but, due to previous significant mass loss, no H-shell burning in their thin hydrogen shells. Their masses cluster around $0.5\,M_\odot$. Only a small fraction of the sdBs show pulsational variations, with non-pulsators also populating the region in the HRD where the pulsators are found. There are p (pressure) mode and g (gravity) mode types of pulsation. Three objects that show both mode types are referred to as hybrid pulsators and among them is V391 Peg (HS 2201+2610) which has five p modes (Østensen et al. 2001; Silvotti et al. 2002) and one g mode (Lutz et al. 2008, 2009). Silvotti

[1] Data obtained with the Hobby-Eberly Telescope (joint project of U of Texas, Pennsylvania State U, Stanford U, U München, U Göttingen) and the W.M. Keck Observatory (operated by CalTech, U of California, NASA), made possible by the generous financial support of the W.M. Keck Foundation.

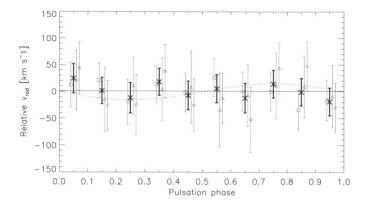

Figure 1: Radial velocities derived for the Hβ (squares), Hγ and Hδ (triangles and diamonds) lines from cross correlation with the template (with error bars, and plotted at a small phase offset for clarity); and weighted mean of the results (crosses and thick error bars) together with a constant and a sinusoidal curve.

et al. (2007) detected parabolic and sinusoidal variations in the observed–calculated (O–C) diagram constructed for the two main pulsation frequencies at 349.5 s and 354.2 s over the observing period of seven years. The sinusoidal component is attributed to the presence of a very low-mass companion (V391 Peg b, $m \sin i = 3.2 \pm 0.7\,M_{\rm Jup}$). The determination of the true mass of this 'asteroseismic planet' requires a constraint on the orbital inclination which can presumably be determined via the stellar rotational inclination.

Limits on the pulsational radial velocities from phase resolved echelle spectra

Echelle spectra of V391 Peg were taken during May and September 2007 with the HRS ($R = 15\,000$) of the HET at the McDonald Observatory, and with HIRES ($R = 31\,000$) at the Keck 1 telescope atop Mauna Kea in May 2008. Data reduction was done with ESO-Midas using standard procedures. Individual echelle orders were merged and the final spectra were carefully normalized and finally summed (Kruspe 2009, diploma thesis, in prep.). This results in a set of individual spectra ($S/N \approx 3$), in particular two September 2007 high time resolution series, and summed spectra for May and September 2007.

In our attempt to "clean" the relevant rotational broadening from pulsational effects, the spectra in September obtained in time resolved mode were combined to a series of ten phase resolved averaged spectra ($S/N \approx 9$) for the main pulsation period of 349.5 s (similar to Tillich et al. 2007).

The cross-correlation of this series of averaged spectra with a pure hydrogen NLTE model spectrum at $T_{\rm eff} = 30\,000$ K and $\log(g/{\rm cm\,s^{-2}}) = 5.5$ as a template yields pulsational radial velocity measurements as shown in Fig. 1 for the different Balmer lines. The maximum amplitude of a sinusoidal curve (fixed at the expected period) that could be accommodated in comparison to the weighted means of the Balmer lines reveals that any pulsational radial velocity amplitude is smaller than the accuracy of our measurements and confirms the upper limit of 16 km s^{-1} given by Kruspe et al. (2008).

The resolution of the model template matches the spectral resolution of the (pulsation-averaged) Keck spectrum. A comparison of the Hα NLTE line core shape yields an even more stringent upper limit for the combined broadening effect of pulsation and rotation of at most 9 km s^{-1}. This means that much better data in terms of spectral resolution and signal to noise will be necessary to measure v_{puls} and $v_{rot} \sin i$.

Acknowledgments. The authors thank A. Reiners and G. Basri for providing the Keck spectrum, H. Edelmann for assistance in obtaining the HET spectra, and U. Heber and T. Rauch for providing grids of model spectra. This work has benefited from the help, advice and software by I. Traulsen. We also thank the conference sponsors and in particular HELAS (European Helio- and Asteroseismology Network, an European initiative funded by the European Commission since April 1st, 2006, as a "Co-ordination Action" under its Sixth Framework Programme FP6) and the Astronomische Gesellschaft for financially supporting the poster presentation at JENAM 2008 Minisymposium N° 4 through travel grants to S.S. and R.L.

References

Kruspe, R., Schuh, S., Silvotti, R., & Traulsen, I. 2008, CoAst, 157, 325

Lutz, R., Schuh, S., Silvotti, R., et al. 2008, ASP Conf. Ser., 392, 339

Lutz, R., Schuh, S., Silvotti, R., et al. 2009, A&A, 496, 469

Østensen, R., Solheim, J.-E., Heber, U., et al. 2001, A&A, 368, 175

Silvotti, R., Janulis, R., Schuh, S., et al. 2002, A&A, 389, 180

Silvotti, R., Schuh, S., Janulis, R., et al. 2007, Nature, 449, 189

Tillich, A., Heber, U., O'Toole, S. J., et al. 2007, A&A, 473, 219

Comm. in Asteroseismology,
Vol. 159, 2009, JENAM 2008 Symposium № 4: Asteroseismology and Stellar Evolution
S. Schuh & G. Handler

Long-term EXOTIME photometry and follow-up spectroscopy of the sdB pulsator HS 0702+6043

R. Lutz,[1,2] S. Schuh,[1] R. Silvotti,[3] R. Kruspe,[1] and S. Dreizler[1]

[1] Institut für Astrophysik, Friedrich-Hund-Platz 1, 37077 Göttingen, Germany
[2] MPI für Sonnensystemforschung, Max-Planck-Straße 2, 37191 Katlenburg-Lindau, Germany
[3] INAF - Osservatorio Astronomico di Capodimonte, via Moiariello 16, 80131 Napoli, Italy

Abstract

Pulsating subdwarf B (sdB) stars oscillate in short-period p modes or long-period g modes. HS 0702+6043 (DW Lyn) is one of a few objects to show characteristics of both types and is hence classified as a hybrid pulsator. It is one of our targets in the EXOTIME program to search for planetary companions around extreme horizontal branch objects. In addition to the standard exercise in asteroseismology to probe the instantaneous inner structure of a star, measured changes in the pulsation frequencies as derived from an O–C diagram can be compared to theoretical evolutionary time scales. Based on the photometric data available so far, we are able to derive a high-resolution frequency spectrum and to report our efforts to construct a multi-season O–C diagram. Additionally, we have gathered time-resolved spectroscopic data in order to constrain stellar parameters and to derive mode parameters as well as radial and rotational velocities.

Individual Objects: HS 0702+6043, HS 2201+2610, HW Vir

Timing method and the EXOTIME program

The O–C (Observed minus Calculated) analysis is a tool to measure the phase variations of a periodic function. The observed times of the pulsation maxima of single runs are compared to the calculated mean ephemeris of the whole data set. Since a low-mass companion, due to its gravitational influence, would cause cyclically advanced and delayed timings of an oscillating sdB's pulsation maxima (due to motion around the common barycentre), this method can be used to search for exoplanets. A sinusoidal component in an O–C diagram is therefore a signature of a companion. In addition, this method can be used to derive evolutionary time scales by measuring linear changes in the pulsation periods. The signature would in this case be a parabolic shape in the O–C diagram. Using this timing method, Silvotti et al. (2007) detected a giant planet companion to the pulsating subdwarf B star HS 2201+2610 (V391 Peg) and recently Lee et al. (2008) reported two planets around the eclipsing sdB+M binary system HW Vir, also revealed by an O–C diagram analysis by measuring the timings of the eclipse minima.

The EXOplanet search with the TIming MEthod (EXOTIME) program is an internationally coordinated effort to examine pulsating sdB stars in terms of planetary companions and evolutionary aspects. Closely related to the puzzling evolution of sdB stars is the late-stage or post-RG evolution of planetary systems and the question if planets could be responsible for the extreme mass loss of the sdB progenitors (e.g. Soker 1998). EXOTIME performs ground based time series photometry from various sites with telescopes in the 0.5 m to 3.6 m range.

Table 1: Our current photometric data archive for HS 0702+6043. Sites: Calar Alto 2.2 m/1.2 m (CA2/1), NOT 2.56 m (N), Göttingen 0.5 m (G), Tübingen 0.8 m (T), Steward Bok 2.2 m (SB), Loiano 1.5 m (L), Mt. Bigelow 1.55 m (MB), Konkoly 1 m (K).

Date		Site	Length [h]	Date		Site	Length [h]
December	1999	CA1	8.4	Nov'07-Mar	2008	MB	424.0
October	2000	N	0.7	March	2008	G	0.6
February	2004	T	7.3	March	2008	L	8.0
February	2004	SB	12.0	March	2008	K	11.5
January	2005	CA2	56.0	May	2008	G	4.1
December	2007	T	31.8	September	2008	G	3.3
December	2007	G	12.0	October	2008	K	2.6
February	2008	T	20.1	October	2008	CA2	4.2
February	2008	G	32.2	November	2008	CA2	8.7

Long-term photometry

HS 0702+6043 was first identified as a variable by Dreizler et al. (2002). It is placed at the common boundary of the p- and g-mode instability regions in a $\log g$-T_{eff} diagram. The two strongest p-mode pulsations at 363.11 s and 383.73 s (amplitudes of 30 and 6 mmag, respectively) will be used to construct multi-seasonal O–C diagrams, for which a time-base of several years is needed. For deriving a single O–C point, at least three to four consecutive nights of observations are needed to provide a sufficient frequency resolution. We aim for a minimum of six O–C points per year. Our data archive for HS 0702+6043 dates back to 1999, unfortunately with large gaps in between. Table 1 lists our current photometric data archive, not yet sufficient to present a meaningful O–C diagram due to the large gap between the observations in 1999 and 2004.

Follow-up spectroscopy

The 772 time resolved high-resolution Echelle spectra (20 s each) of HS 0702+6043 taken at the Hobby Eberly Telescope will provide rotational and in particular pulsational radial velocities. The pulsational amplitudes are expected to be larger for HS 0702+6043 than for HS 2201+2610 since the photometric amplitudes are larger. An analysis of the time-resolved spectroscopy of HS 2201+2610 can be found in Schuh et al. (2009).

Acknowledgments. The authors thank all observers who contributed observations to the HS 0702+6043 data archive: B. Beeck, Z. Bognar, E. M. Green and collaborators, M. Hundertmark, T. Nagel, R. Østensen, M. Paparo, P. Papics, T. Stahn. Partly based on observations collected in service mode by L. Montoya, M. Alises and U. Thiele for our program H08-2.2-009 at the Centro Astronómico Hispano Alemán (CAHA) at Calar Alto, operated jointly by the Max-Planck Institut für Astronomie and the Instituto de Astrofísica de Andalucía (CSIC). The authors thank the Astronomische Gesellschaft as well as the conference sponsors and in particular HELAS (European Helio- and Asteroseismology Network, an European initiative funded by the European Commission since April 1st, 2006, as a "Co-ordination Action" under its Sixth Framework Programme FP6) for financially supporting the poster presentation at JENAM 2008 Minisymposium N° 4 through travel grants to RL and SS.

References

Dreizler, S., Schuh, S., Deetjen, J. L., et al. 2002, A&A, 386, 249

Lee, J. W., Kim, S.-L., Kim, C.-H., et al. 2008, arXiv:0811.3807

Schuh, S., Kruspe, R., Lutz, R., & Silvotti, R. 2009, CoAst, 159, 91

Silvotti, R., Schuh, S., Janulis, R., et al. 2007, Nature, 449, 189

Soker, N. 1998, AJ, 116, 1308

Comm. in Asteroseismology,
Vol. 159, 2009, JENAM 2008 Symposium № 4: Asteroseismology and Stellar Evolution
S. Schuh & G. Handler

Search for sdB/WD pulsators in the Kepler FOV

R. Silvotti,[1] G. Handler,[2] S. Schuh,[3] B. Castanheira,[2] and H. Kjeldsen[4]

[1] INAF – Osservatorio Astronomico di Capodimonte, via Moiariello 16, 80131 Napoli, Italy
[2] Institut für Astronomie, Universität Wien, Türkenschanzstrasse 17, 1180 Vienna, Austria
[3] Institut für Astrophysik, Universität Göttingen, Friedrich-Hund-Platz 1, 37077 Göttingen, Germany
[4] Dept. of Physics and Astronomy, Aarhus University, Ny Munkegade, 8000 Aarhus C, Denmark

Abstract

In this article we present the preliminary results of an observational search for subdwarf B and white dwarf pulsators in the Kepler field of view (FOV), performed using the DOLORES camera attached to the 3.6 m *Telescopio Nazionale Galileo* (TNG).

Individual Objects: KIC10_05807616, KIC10_02020175

Introduction

The Kepler satellite will be launched in March 2009 and will observe a $\sim$105 square degree field for 4 years with the primary goal of finding new exoplanets using the transit method. Kepler's secondary goal is asteroseismology: the objective is to characterize stars hosting planets, and also to study in detail a few thousands other oscillating stars. Among the seismic targets, up to 512 stars can be observed in short cadence with a sampling time of 1 minute (for all the other targets the cadence will be 30 min), allowing the study of short period pulsators, including hot subdwarfs B (sdBs) and white dwarfs (WDs). Thanks to its exceptional photometric accuracy and duty cycle ($\approx$95%, see Christensen-Dalsgaard et al. 2006 for more details), Kepler can produce numerous exciting results on these stars: 1) detect low-amplitude ($\lesssim$100 ppm) and high-degree (l>2) modes, not visible from the ground. 2) Measure stellar global parameters with unprecedented accuracy (mass, rotation, H/He layer thickness, T_{eff}, $\log g$). 3) Improve our understanding of the physics of these stars (differential rotation; core C/O ratio and equation of state, neutrino cooling and crystallization in WDs). 4) Study amplitude variations and nonlinear effects. 5) Through the O–C diagram, measure $\dot{P}$, determine the evolutionary status of the star and search for low-mass companions (BDs/planets) with masses down to $\approx$10^{-1}M$_{Jup}$ (see the recent example of V391 Peg b, Silvotti et al. 2007).

Observations and preliminary results

The 24 targets were selected from the KIC10 (Kepler Input Catalogue version 10, used internally by the Kepler team to select targets) through their $g - i$ SLOAN colour. For most of the targets proper motions were available from the USNO catalogue allowing to refine the selection using a reduced proper motion diagram.

The time-series photometry was performed during a single run at the 3.6 m TNG in August 2008. Each target was observed for 1 to 2 hours with the SLOAN g filter, with exposure

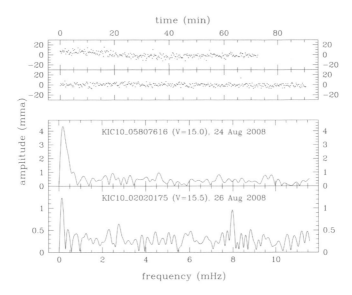

Figure 1: Light curve and amplitude spectrum of the two best pulsator candidates (see text for more details).

times between 1 and 10 s, depending on the magnitude. Details on the selection criteria and data reduction will be given in a forthcoming paper (Silvotti et al., in preparation).

A new variable star was found from a preliminary analysis of our data. It is probably a cataclysmic variable and we will present it in a future article. However, our data did not revealed any star with a clear signature of intrinsic pulsations. The typical upper limits that we have obtained for the pulsation amplitude are between 1 and 2 mma (1 mma = milli-modulation-amplitude = 1000 ppm). Nevertheless, for two targets having colours compatible with sdB stars, the light curve or the amplitude spectrum suggest possible periodicities. The upper panel target in Fig. 1 shows a periodicity near 1 hour, compatible with a slow (g mode) sdB pulsator. The lower panel target has a peak higher than 3 times the local noise (S/N>3) at about 125 s, which would correspond to a rapid (p mode) sdB pulsator.

New observations of these two stars done in October 2008, and now under reduction, will help to clarify whether they oscillate or not. Just before submitting this article, we have been informed that all the 24 targets observed at the TNG have been included in a list of *survey targets* that should be observed by Kepler in the first months of the mission to verify their pulsational stability with a much lower detection threshold.

Acknowledgments.   The authors are very grateful to Alfio Bonanno and Silvio Leccia for having given part of their six TNG nights to this programme, following technical problems with the SARG instrument. Silvio Leccia has been also observing during the first night. RS wishes to thank the TNG technical team, in particular Gloria Andreuzzi, Antonio Magazzu and Luca Difabrizio, who did an excellent job during the 7 nights of service-mode observations.

References

Christensen-Dalsgaard, J., Arentoft, T., Brown, T. M., et al. 2006, CoAst, 150, 350

Silvotti, R., Schuh, S., Janulis, R., et al. 2007, Nature, 449, 189

Comm. in Asteroseismology,
Vol. 159, 2009, JENAM 2008 Symposium № 4: Asteroseismology and Stellar Evolution
S. Schuh & G. Handler

Asteroseismology and evolution of GW Vir stars

P.-O. Quirion

Aarhus Universitet, Institut for Fysik og Astronomi, Ny Munkegade Bld. 1520, DK-8000 Aarhus, Danmark

Abstract

We present a summary of current knowledge of the hot, pulsating, hydrogen deficient GW Vir stars. At $T_{eff} \simeq 120,000$ K and with an atmosphere composed mainly of helium, carbon, and oxygen, these pulsating stars are showing multiperiodic, intermediate- to high-order g modes. They include in their class some of the hottest white dwarfs and planetary nebulae. We present non-adiabatic asteroseismology results used to determine the atmospheric parameters of GW Vir stars. We also discuss some of the latest asteroseismology results using state of the art post-Asymptotic Giant Branch evolution models.

Individual Objects: Longmore 4, Abell 43, NGC 7094, NGC 246, NGC 5189, Sk 3, NGC 2867, NGC 6905, NGC 1501, HE 1429−1209, RX J2117.1+3412, HS 2324+3944, NGC 2371, K1-16, PG 1159−035, VV 47, PG 2131+066, PG 1707+427, PG 0122+200, H1504+65

Introduction

From $T_{eff} \sim 85\,000$ K to $200\,000$ K and from $\log g = 5.5$ to 8.0 extends a region of the $\log g - T_{eff}$ diagram seated exactly on the "knee" leading from the post-Asymptotic Giant Branch (post-AGB) to the white dwarf cooling track. This region is filled with a variety of hydrogen deficient stars being referred to, more and more in recent articles and conferences, as pre-white dwarfs. A multitude of spectral and sub-spectral types are included under that generic name. We can regroup them under three main types. The PG 1159 stars, the Early type Wolf Rayet Central Stars of Planetary Nebulae, or more conveniently the [WCE] stars, and the O(He) stars.

The [WCE] and PG 1159 stars are both believed to be the products of the born-again scenario in which hydrogen rich stars entering the white dwarf cooling track are pushed back to the AGB by a (very)-late thermal pulse (Iben et al. 1983, Herwig et al. 1999). At the end of the born-again process, the star moves back toward the white dwarf branch and is highly or completely depleted of hydrogen. At this point, it shows a significant amount of helium, carbon and oxygen in its spectra. The range of abundance is $X(He) \simeq 0.30 - 0.85$, $X(C) \simeq 0.15 - 0.60$, and $X(O) \simeq 0.02 - 0.20$, by mass. There is also the peculiar object H1504+65 with its nearly pure carbon and oxygen spectra $X(C) \simeq X(O) \simeq 0.50$, and the so called hybrid stars which show a relatively large amount of hydrogen, $X(H) \simeq 0.15 - 0.35$. More details about abundance patterns and the evolutionary history of these stars are found in Werner & Herwig (2006). The origin of O(He) stars is somewhat more nebulous. They show high helium enrichment with only traces of the CNO elements in their spectra with composition of $X(H) \simeq 0.10 - 0.50$ $X(He) \simeq 0.50 - 0.90$. We refer the reader to Rauch et al. (2008) and references therein for more details. In any case, the fate of all pre-white dwarfs is to become white dwarfs. Those that are completely hydrogen depleted should turn into

Table 1: Confirmed GW Vir stars.

Star	log g	$T_{eff}[kK]$	Periods [s]	Remark
Longmore 4	5.5	120	831−2325	transient
Abell 43	5.7	110	2604−5529	hybrid
NGC 7094	5.7	110	2000−5000	hybrid
NGC 246	5.7	150	480−4560	
NGC 5189	6.0	135	690	[WCE]
Sk 3	6.0	140	929−2183	[WCE]
NGC 2867	6.0	141	769	[WCE]
NGC 6905	6.0	141	710−912	[WCE]
NGC 1501	6.0	134	1154−5235	[WCE]
HE 1429−1209	6.0	160	919	
RX J2117.1+3412	6.0	170	694−1530	
HS 2324+3944	6.2	130	2005−2569	hybrid
NGC 2371	6.3	135	923−1825	[WC]-PG 1159
K1-16	6.4	140	1500−1700	
PG 1159−035	7.0	140	339−982	
VV 47	7.0	130	261−4310	
PG 2131+066	7.5	95	339−508	
PG 1707+427	7.5	85	336−942	
PG 0122+200	7.5	80	336−612	

helium rich DO white dwarfs, while the others, with various amounts of hydrogen, should transform into DA white dwarfs.

To this description of the pre-white dwarf region, we add the main topic of this review, which is the GW Vir instability strip. The strip covers a large part of the pre-white dwarf region and is caused by a classical κ mechanism triggered by partial ionization of the K-shell of both carbon and oxygen (Starrfield et al. 1983, Starrfield et al. 1984, Cox 2003, Quirion et al. 2007a). We have observed 19 [WCE] and PG 1159 stars in the GW Vir instability strip. These pulsating stars are showing medium to high radial order g modes with $k \gtrsim 15$ and $l = 1, 2$. None of the O(He) stars is known to be variable. Only small traces of carbon are detected in the spectra of one of the four O(He) stars, not enough to trigger oscillations and make it part of the GW Vir instability strip. Table 1 presents an updated list of the GW Vir stars with their stellar parameters and detected instability range. References for this table can be found in Quirion et al (2007a), except for VV 47 (González Pérez et al. 2006) and NGC 7094 (Solheim et al. 2007) which have been discovered to pulsate only recently.

Non-adiabatic studies

In Quirion et al. (2007a), the position of the blue edge of the GW Vir instability strip is shown to be a function of the exact carbon and oxygen content in the envelope of the pre-white dwarf. A rule of thumb is to state that the higher the content of carbon and oxygen in the pre-white dwarf is, the higher the temperature of its blue edge will be. Since pre-white dwarfs are showing a large range of chemical composition, there is accordingly a large set of possible blue edges. This phenomenon is depicted in Fig. 1. Following that simple idea, two reasons are invoked to explain the stability of a star located in the pre-white dwarf region. First, the depletion of C and O in the envelope of the star, which makes the driving of pulsation impossible, and the high temperature of the star that puts it outside the instability strip. The later reason is invoked to explain why H1504+65, a pure C-O but extremely hot (200 000 K) star, is stable.

The stability of pre-white dwarfs is thus determined by the chemical composition of the envelope and the effective temperature. Since these two parameters are fixed by spectroscopy,

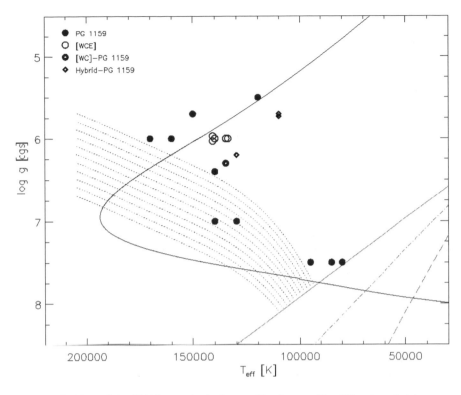

Figure 1: Positions of the GW Vir stars in the $\log g - T_{\text{eff}}$ diagram. The different spectral types are specified by different symbols. The blue dotted curves symbolize the intrinsic fuzziness of the blue edge caused by its dependence on the atmospheric chemical composition. The red lines are the theoretical red edges calculated with full evolutionary models including diffusion and mass loss. Each red edge is calculated with a different mass loss law, $\dot{M} = 1.00 \times 10^{-17} L^{2.36}$ refers to the solid red line, $\dot{M} = 1.29 \times 10^{-15} L^{1.86}$ refers to the dot-dashed red line and $\dot{M} = 1.82 \times 10^{-13} L^{1.36}$ refers to the dashed red line.

one can check the consistency of the spectroscopic measurement by comparing the observed instability range of GW Vir stars with the period range obtained with non-adiabatic computations. Table 2 shows the results of that exercise for a sample of GW Vir stars. One can compare the unstable period ranges obtained by models using the spectroscopic values for the composition and the atmospheric parameters of Table 2 to the observed ranges of Table 1 and state that the matches are qualitatively good. However, we can find better fits to the observed period bands by varying the gravity and effective temperature of the models within observational uncertainties. The results of this exercise, explained in more details in Quirion et al. (2008, 2009) is shown in Table 2 under "Non-adiabatic asteroseismology". An explicit computation is shown in Fig. 2 for PG 1159−035, the prototype of the class and the best studied GW Vir star. Non-adiabatic asteroseismology can be considered as a consistency check of spectroscopic measurements. Nonetheless, as Table 2 shows, non-adiabatic asteroseismology also gives a significant improvement to the GW Vir stars' measured surface gravity, and to some extent to their measured effective temperature.

Table 2: Non-adiabatic asteroseismology.

Star	Quantitative spectroscopy			Non-adiabatic asteroseismology		
	$\log g$ ±.5 dex	T_{eff} [kK] ±∼10%	Periods [s]	$\log g$	l_{eff} [kK]	Periods [s]
NGC 246	5.70	150	514−11453	∼5.75	−	604−4477
RX J2117.1+3412	6.00	170	635−2202	∼6.10	∼180±10	681−1530
PG 1159−035	7.00	140	247−623	6.80±.05	144± 3	336−987
VV 47	7.00	130	216−891	6.10±.10	130±20	235−3531
PG 2131+066	7.50	95	197−650	7.25±.25	−	118−508
PG 1707+427	7.50	85	188−765	7.35±.05	81± 5	224−960
PG 0122+200	7.50	80	198−847	7.50±.30	−	330−602

Red edge

The main factor that differentiates PG 1159 from the [WCE] stars is the presence of a stronger radiatively driven wind in the latter. The wind produces wide strong C IV, He II and O VI emission lines, giving the [WCE] type its characteristic spectral signature. With time and decreasing luminosity, the magnitude of the outgoing wind diminishes, thus weakening the carbon emission lines in the [WCE] stars and letting the PG 1159 features appear in the spectra of the [WC]-PG 1159 transition objects. A further decrease of the mass loss and wind completes the scenario whereby the wide carbon emission lines disappear, leading to a full characteristic PG 1159 spectrum. Hence the sequence [WCE]→PG 1159.

The following evolutionary step connects PG 1159 to DO stars, where the spectral type DO belongs to the He-dominated atmosphere white dwarfs. This scenario is also strongly supported by quantitative spectroscopy (Werner & Herwig 2006). The strong wind present in [WC] and PG 1159 stars tends to homogenize their envelope (see, e.g. Chayer et al. 1997, Unglaub & Bues 2001). As this wind weakens with lowering luminosity, it progressively looses its homogenizing capacity and gravitational settling takes over. This will cause carbon and oxygen to precipitate while helium will float at the surface of the star. As carbon and oxygen are responsible for the κ mechanism in these stars, it is expected that gravitational settling will ultimately bring the driving to a stop, drawing the red edge of the class (Quirion et al. 2007b). The temperature of the red edge is linked to the strength of the mass loss. A weaker mass loss will make the carbon and oxygen sink more rapidly while the stronger wind will keep these elements afloat a longer time and push the red edge to lower temperature. Results from Quirion et al. (2007b), showing the effect of the mass loss strength on the position of the red edge, using full evolutionary calculation including diffusion and mass loss, are pictured in Fig. 1.

Adiabatic studies

Non-adiabatic processes can be studied with only an approximate knowledge of GW Vir cores. These processes happen above $\log q = \log(1 - m(r)/M_\star) \lesssim -6$ and only this region will affect the instability range of these stars. However, GW Vir asteroseismology, which does a one by one comparison between observed modes and those of putative models, is more influenced by the composition and structure below $\log q \gtrsim -6$. As shown in Fig. 8 of the pulsating white dwarf review by Fontaine & Brassard (2008), the importance of the deeper layers in the construction of a star's eigenmodes increases with temperature on the white dwarf track. The chemical stratification of the core will, unlike the case of DBV and DAV white dwarfs, play an important role in the construction of the GW Vir eigenmodes.

Another factor influencing the structure of GW Vir stars, especially those at low gravity,

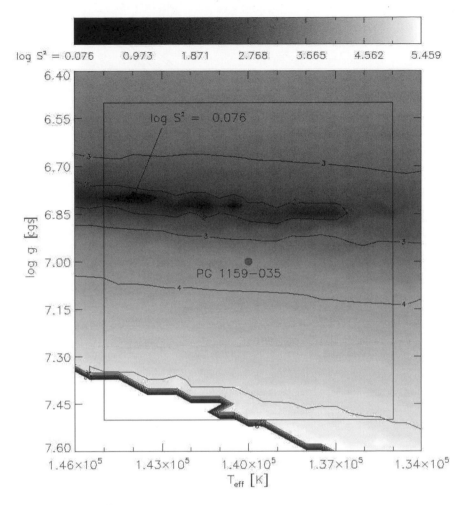

Figure 2: Calculated S^2 function grid (in logarithmic units) for PG 1159−035 in the $\log g - T_{eff}$ diagram. The spectroscopy position of the star is shown with a red dot. The box represents its associated uncertainties. The arrow points to the best-fit non-adiabatic model ($\log g = 6.8$ and $T_{eff} = 144\,000$ K). The S^2 function used for the fit is described in detail in Quirion et al. (2008, 2009).

is their memory of the thermal pulse undergone during the born-again phase. This means that the region where the thermal pulse occurred, around $\log q \sim -2$, can be significantly far from thermal equilibrium. For that reason, the exact treatment of mixing and burning should influence the path of a model in the pre-white dwarf region, and will certainly influence the model eigenmodes (Herwig & Austin 2004, Miller Bertolami & Althaus 2007).

Presently, the most sophisticated and extended set of models evolving through the pre-white dwarf region is the one of the La Plata group (Althaus et al. 2005, Miller Bertolami & Althaus 2006). Their models are starting on the main sequence and ultimately pass through

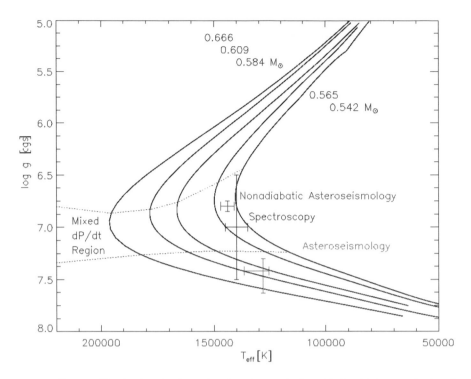

Figure 3: Five pre-white dwarf tracks after undergoing a born-again phase. Enclosed in between the dotted lines are the models showing simultaneously positive and negative (mixed) values of $\dot{P}_k$ (Miller Bertolami & Althaus 2006). Along the tracks are three fits for the star PG 1159−035. The asteroseismology fit (Córsico et al. 2008) is at higher gravity and derived directly from the presented tracks, the spectroscopic fit (Jahn et al. 2007) from non-LTE atmosphere modelling, and the non-adiabatic fit from static envelope models using the chemical composition derived from the spectroscopic fitting (Quirion et al. 2008).

the born-again phase before they reach the pre-white dwarf region. Some of their tracks are reproduced in Fig. 3. An interesting characteristic of the seismology of these tracks is the region where the temporal variation of the period of individual modes $\dot{P}_k$ is positive for some of the modes k and negatives for others. This mixed $\dot{P}$ region marks the transition between the post-AGB, where gravitational contraction dominates and $\dot{P}_k \sim -d(\log g)/dt \lesssim 0$, $\forall k$, and the white-dwarf track, where cooling dominates and $\dot{P}_k \sim -dT_{\mathrm{eff}}/dt \gtrsim 0$, $\forall k$.

It its quite remarkable that over 30 years of measurements, we have been able to determine with increasing precision the value of $\dot{P}$ for several modes of PG 1159−035. Even more remarkable is the mixed $\dot{P}$ behavior of its modes. Table 3 shows all measurements with $l = 0$ and $m = 0$ of $\dot{P}_k$ where k is the radial order taken from (Costa & Kepler 2008).

As mentioned in the caption of Fig. 3, fits for PG 1159−035 from spectroscopy, non-adiabatic asteroseismology, and asteroseismology are superimposed to the tracks. The asteroseismology fit derives directly from the presented tracks (Córsico et al. 2008). Unfortunately, the adiabatic fit is not able to reproduce the mixed $\dot{P}$ behavior of the star. Also, even if they are reasonably close, it seems that spectroscopy and non-adiabatic asteroseismology, which both depend on the structure of the upper part of the star, are at odds with the core-influenced fit of asteroseismology.

Table 3: $\dot{P}$ rate measurements for the modes of PG 1159−035 identified as $l = 1$ and $m = 0$.

P [s]	k	$\dot{P} \times 10^{-1}$
390.3	14	10.4±2.3
452.4	17	2.6±0.5
494.8	19	−30.1±2.8
517.1	20	18.2±0.8
538.1	21	1.0±0.7
558.4	22	−4.3±1.9

The reason of the disparity between the asteroseismological fit and the other aspects of the problem probably rests on the following factors. PG 1159−035 shows medium- to high-order modes with $k \gtrsim 15$. This means that the modes are close to the asymptotic regime where the periods and their spacing depend more on the star's global parameters (Unno et al. 1989) and less on the exact structure of the star. It is therefore reasonable to suggest that a family of solutions and not a single model should be able to yield good asteroseismological fits to PG 1159−035. One could picture this assessment of degenerate solutions by comparing the fits to PG 1159−035 coming from two very different generations of models; the older study of Kawaler & Bradley (1994) and the modern study of Córsico at al. (2008). Both studies give comparably good asteroseismological fits to the star's eigenfrequencies. We can consider each of the model sets to be a slice of the pre-white dwarf parameter space and state that each of the fits is a solution in the valley of good solutions for PG 1159−035. These are only two studies and the parameter space for asteroseismology (core composition, effect of the last thermal pulse, etc.) is still to be explored in detail. Unfortunately, the construction of modern pre-white dwarf models involves heavy computations.

There is a brighter side to this puzzle. We have in sight the deeper structure of GW Vir stars. One might achieve the goal of understanding these deeper layers by looking at every one of the multiple angles revealed by GW Vir stars, spectroscopy, non-adiabatic processes, asteroseismology, and $\dot{P}$, and try to build a set of models that would fit all these aspects at the same time. From there, we would answer the central problem of stellar astrophysics, which is to know a star from its surface to its core.

Acknowledgments. I would like to thank G. Fontaine for his interest in this work.

References

Althaus, L. G., Serenelli, A. M., Panei, J. A., et al. 2005, A&A, 435, 631

Bond, H. E. 2008, ASP Conf. Ser., 391, 129

Chayer, P., Fontaine, G., & Pelletier, C. 1997, ASSL, 214, 253

Córsico, A. H., Althaus, L. G., Kepler, S. O., et al. 2008 A&A, 478, 869

Costa, J. E. S., & Kepler, S. O. 2008, A&A 489, 1225

Cox, A. N. 2003, ApJ, 585, 975

Fontaine, G., & Brassard, P. 2008, PASP, 120, 1043

González Pérez, J. M., Solheim, J.-E., & Kamben, R. 2006, A&A, 454, 527

Herwig, F., Blöcker, T., Langer, N., & Driebe, T. 1999, A&A, 349, L5

Herwig, F., & Austin, S. M. 2004, ApJ, 613, L73

Iben, I., Jr., Kaler, J. B., Truran, J. W., & Renzini, A. 1983, ApJ, 264, 605

Jahn, D., Rauch, T., Reiff, E., et al. 2007, A&A, 426, 281

Kawaler, S. D. & Bradley, P. A. 1994, ApJ, 427, 415

Miller Bertolami, M. M., & Althaus, L. G. 2006, A&A, 454, 845

Miller Bertolami, M. M., & Althaus, L. G. 2007, A&A, 470, 675

Quirion, P.-O., Fontaine, G., & Brassard, P. 2007a, ApJS, 171, 219

Quirion, P.-O., Fontaine, G., & Brassard, P. 2007b, ASP Conf. Ser., 372, 649

Quirion, P.-O., Fontaine, G., & Brassard, P. 2008, J. Phys.: Conf. Ser., 118, 012069

Quirion, P.-O., Fontaine, G., & Brassard, P. 2009, J. Phys.: Conf. Ser., in press

Rauch, T., Reiff, E., Werner, K., & Kruk, J. W. 2008, ASP Conf. Ser., 391, 135

Solheim, J.-E., Vauclair, G., Mukadam, A. S., et al. 2007, A&A, 468, 1057

Starrfield, S. G., Cox, A. N., Hodson, S. W., & Pesnell, W. D. 1983, ApJ, 268, L27

Starrfield, S., Cox, A. N., Kidman, R. B., & Pesnell, W. D. 1984, ApJ, 281, 800

Unglaub, K., & Bues, I. 2001, A&A, 374, 570

Unno, W., Osaki, Y., & Ando, H. 1989, in "Nonradial oscillations of stars", University of Tokyo Press

Werner, K., Hamann, W.-R., Heber, U., et al. 1992, A&A, 259, L69

Werner, K., & Herwig, F. 2006, PASP, 118, 183

Comm. in Asteroseismology,
Vol. 159, 2009, JENAM 2008 Symposium № 4: Asteroseismology and Stellar Evolution
S. Schuh & G. Handler

UV spectroscopy of the hybrid PG 1159-type central stars of the planetary nebulae NGC 7094 and Abell 43

M. Ziegler,[1] T. Rauch,[1] K. Werner,[1] L. Koesterke,[2] and J. W. Kruk[3]

[1] Institute for Astronomy and Astrophysics, Kepler Center for Astro and Particle Physics,
Eberhard Karls University, Tübingen, Germany
[2] Texas Advanced Computing Center, University of Texas, Austin, U.S.A.
[3] Department of Physics and Astronomy, Johns Hopkins University, Baltimore, U.S.A.

Abstract

Hydrogen-deficient post-AGB stars have experienced a late helium-shell flash that mixes the hydrogen-rich envelope and the helium-rich intershell. The amount of hydrogen remaining in the stellar envelope depends on the particular moment when this late thermal pulse occurs.

Previous spectral analyses of hydrogen-deficient post-AGB stars revealed strong iron deficiencies of up to 1 dex. A possible explanation may be neutron captures due to an efficient s-process on the AGB that transformed iron into heavier elements. An enhanced nickel abundance would, thus, be an indication for this scenario.

We performed a detailed spectral analysis by means of NLTE model-atmosphere techniques based on high-resolution UV observations of the two PG 1159-type central stars of the planetary nebulae NGC 7094 and Abell 43 which are spectroscopic twins, i.e. they exhibit very similar spectra. We confirmed a strong iron-deficiency of at least one dex in both stars. The search for nickel lines in their UV spectra was entirely negative. We find that both stars are also nickel-deficient by at least one dex.

Individual Objects: NGC 7094, Abell 43

Introduction

NGC 7094 and Abell 43 are so-called hybrid-PG 1159 type central stars of planetary nebulae. In contrast to other PG 1159 stars they exhibit hydrogen lines in their spectra. Hybrid-PG 1159 stars experience a final thermal pulse (TP) while they are still on the asymptotic giant branch (AGB) and hydrogen-shell burning is still "on" (cf. Herwig 2001). Hydrogen is then diluted to $\sim 17\%$ by mass.

In a previous spectral analysis, Dreizler et al. (1997) found $T_{\rm eff} = 110$, $\log g = 5.7$, and H:He $= 36{:}43$ (by mass) for both stars, NGC 7094 and Abell 43. A determination of the iron abundance of NGC 7094 revealed a strong deficiency of at least one dex (Miksa et al. 2002). Neutron capture due to the s-process in the former AGB star could have transformed Fe into heavier nuclei.

At the relevant temperature regime, all elements are highly ionized. The dominant ionization stages of the iron-group elements in the line-forming regions are VI and VII. Lines of these are located in the UV and FUV range, which were only accessible by the Far Ultraviolet Spectroscopic Explorer (FUSE) and the Hubble Space Telescope (HST) with STIS (Space Telescope Imaging Spectrograph). To determine the iron and nickel abundances of

NGC 7094 and Abell 43, we analyzed high-resolution and high S/N observations obtained with these spectrographs.

Spectral analysis

Since the spectral analysis of NGC 7094 and Abell 43 by Dreizler et al. (1997) using low resolution TWIN spectra, our model atmospheres as well as the available observations have been improved (see Rauch & Deetjen 2003, Werner et al. 2003 for details). Based on a high-resolution ($R = 18\,000$), high-S/N optical spectrum (obtained during the SPY campaign, Napiwotzki et al. 2001) and theoretical spectral energy distributions (SEDs) provided by *TheoSSA*[1], a service by *GAVO*[2], we found that we achieve a much better fit with the observation at $T_{eff} = 100 \pm 10\,$kK, $\log g = 5.0 \pm 0.3$, and H:He $= 9{:}79$ ($\pm 0.5\,$dex). Due to a lower S/N, the error ranges are a factor of two larger for Abell 43.

We adopted these parameters and calculated plane-parallel, hydrostatic NLTE model atmospheres using the Tübingen Model Atmosphere Package (*TMAP*), which considered the opacities of 23 elements (H, He, C, N, O, F, Ne, Na, Mg, Al, Si, P, S, Ar, Ca, Sc, Ti, V, Cr, Mn, Fe, Co, and Ni). The abundances of C, N, O, and Si could be determined ([X] $= 1.7, -0.9, -0.2, 0.4$, respectively; [X] denotes log (abundance / solar abundance) of species X). For the other elements, only upper limits could be determined (for details, see Ziegler et al. in prep.). The subsolar oxygen abundance is unexplained because carbon as well as oxygen should be dredged up during the thermal pulse and therefore be enhanced.

Our analysis of NGC 7094 and Abell 43 confirmed the strong Fe-deficiency that was previously found (Miksa et al. 2002). Due to the lack of reliable atomic data for post-iron elements and the expected high ionization stages, we were able to search for cobalt and nickel lines only. Although we have derived new parameter values which improved the agreement between model and observation, we were not able to identify any Co or Ni line. Test calculations show that these lines should be visible at [Co] $= 0$ and [Ni] $= -1$. While an increased Ni abundance would be a clear evidence for a transformation of Fe into Ni by neutron captures, the Ni-deficiency may be an indication that the s-process transformed Ni into even heavier elements. Due to the lack of suitable atomic data at the investigated parameter regime, the analysis of abundances of trans-iron group elements is presently impossible.

Results and conclusions

Our analysis of the two hybrid PG 1159 stars NGC 7094 and Abell 43 confirmed the previously found Fe-deficiency of at least one dex. Although we have derived new parameter values which improved the agreement between model and observation, we were not able to find any Ni enhancement. Instead, we found Ni to be deficient by at least one dex. It is therefore possible that the s-process also converted Ni into even heavier trans-iron group elements. Unfortunately, we are presently not able to search for these elements because of the lack of reliable atomic data of the expected high ionization stages.

Acknowledgments. M.Z. thanks the Astronomische Gesellschaft for a travel grant. T.R. is supported by the *German Astrophysical Virtual Observatory* (GAVO) project of the German Federal Ministry of Education and Research (BMBF) under grant 05 AC6VTB. J.W.K. is supported by the FUSE project, funded by NASA contract NAS5—32985.

---

[1] http://vo.ari.uni-heidelberg.de/ssatr-0.01/TrSpectra.jsp?
[2] German Astrophysical Virtual Observatory

References

Dreizler, S., Werner, K., & Heber, U. 1997, in "Planetary nebulae", eds. H. J. Habing and
    H. J. G. L. M. Lamers, IAU Symposium, 180, 103

Herwig, F. 2001, Ap&SS, 275, 15

Miksa, S., Deetjen, J. L., Dreizler, S., et al. 2002, A&A, 389, 953

Napiwotzki, R., Christlieb, N., Drechsel, H., et al. 2001, AN, 322, 411

Rauch, T., & Deetjen, J. L. 2003, in "Stellar Atmosphere Modeling", eds. I. Hubeny, D. Mihalas, and
    K. Werner, ASP Conf. Ser., 288, 103

Werner, K., Deetjen, J. L., Dreizler, S., et al. 2003, in "Stellar Atmosphere Modeling", eds. I. Hubeny,
    D. Mihalas, and K. Werner, ASP Conf. Ser., 288, 31

Poster session and discussions in the Arcades.

Comm. in Asteroseismology,
Vol. 159, 2009, JENAM 2008 Symposium № 4: Asteroseismology and Stellar Evolution
S. Schuh & G. Handler

Towards a dynamical mass of a PG 1159 star: radial velocities and spectral analysis of SDSS J212531−010745

B. Beeck,[1] S. Schuh,[1] T. Nagel,[2] and I. Traulsen[1]

[1] Institut für Astrophysik, Universität Göttingen, Friedrich-Hund-Platz 1, 37077 Göttingen, Germany
[2] Institut für Astronomie und Astrophysik, Universität Tübingen, Sand 1, 72076 Tübingen, Germany

Abstract

The evolutionary scenarios which are commonly accepted for PG 1159 stars are mainly based on numerical simulations, which have to be tested and calibrated with real objects with known stellar parameters. One of the most crucial but also quite uncertain parameters is the stellar mass. PG 1159 stars have masses between 0.5 and 0.8 $M_\odot$, as derived from asteroseismic and spectroscopic determinations. Such mass determinations are, however, themselves model-dependent. Moreover, asteroseismically and spectroscopically determined masses deviate systematically for PG 1159 stars by up to 10%.

SDSS J212531.92-010745.9 is the first known PG 1159 star in a close binary with a late-main-sequence companion allowing a dynamical mass determination. We have obtained 14 Calar Alto spectra of SDSS J212531.92−010745.9 covering the full orbital phase range. A radial velocity curve was extracted for both components. With co-added phase-corrected spectra the spectral analysis of the PG 1159 component was refined. The irradiation of the companion by the PG 1159 star is modelled with PHOENIX, yielding constraints on radii, effective temperature and separation of the system's components. The light curve of SDSS J212531.92−010745.9, obtained during three seasons of photometry with the Göttingen 50 cm and Tübingen 80 cm telescopes, was modelled with both the nightfall and PHOEBE programs.

Individual Objects: SDSS J212531.92-010745.9

Extraction of the radial velocity curves

In August 2007, 14 spectra of SDSS J212531.92-010745.9 covering the total phase range were taken with the TWIN spectrograph at the 3.5m telescope at Calar Alto Observatory (Alméria, Spain). These show typical PG 1159 features together with the Balmer series of hydrogen in emission (plus other emission lines), already interpreted as signature of an irradiated close companion by Nagel et al. (2006) from an SDSS spectrum. The Calar Alto spectra cover wavelengths from 3800 Å up to about 7000 Å. At 5000 Å they have a resolution of $R = 4170$ and a SNR ranging from 4 to 13. The spectra were reduced and normalized to their continua. To determine the radial velocity (RV) curve of the secondary of the system a Gaussian was fitted to seven H Balmer lines (H α, H β, ..., H η, emission line height variable with phase) to locate the line centres. To evade uncertainties in the wavelength calibration and intrinsic wavelength dependence of the RV, a sine was fitted to the RV curve separately for each line. The zero points of the fits were shifted to the mean zero point $\langle v_0 \rangle = 81.5\,\mathrm{km\,s^{-1}}$ with an uncertainty of $\pm 9.4\,\mathrm{km\,s^{-1}}$. The weighted

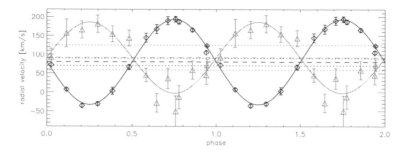

Figure 1: Radial velocities (RVs) derived from the Calar Alto spectra. The grey symbols and error bars signify the RVs of the PG 1159 component together with the sine fit (solid grey line; amplitude $K_1 = (94.3 \pm 15.0)$km s^{-1}), the zero point of which is shown as dashed line. The black symbols signify the RVs of the secondary, again with the sine fit (solid black line; amplitude $K_2 = (113.0 \pm 3.0)$km s^{-1}) and its zero point (dashed black line). The dotted lines are the uncertainties of the zero points.

mean of the shifted RVs was calculated and a sine was fitted to the resulting mean RVs yielding a secondary RV amplitude of $K_2 = (113.0 \pm 3.0)$ km s^{-1}. The primary RV curve was deduced from the cross correlation of a small section of the spectra containing two narrow C IV absorption lines at 5801/5812 Å, which were unblended with the strong emission lines of the companion. The accuracy of the RV values obtained by this method was estimated to amount to ± 30 km s^{-1} and was rescaled with the SNR of the individual spectra. Again a sine was fitted to the RVs. The amplitude of this fit is $K_1 = (94.3 \pm 15.0)$ km s^{-1} and its zero point is at $v_{0,\text{PG 1159}} = (92 \pm 33)$ km s^{-1}, slightly redshifted with respect to the zero point of the RV curve of the secondary. Although this redshift is insignificant, it raises the hope to detect a gravitational redshift for the PG 1159 component of SDSS J212531.92−010745.9 with high-resolution spectroscopy in subsequent studies. The RV curves are shown in Fig. 1.

Spectral analysis and light curve modelling

Using the RV curves obtained, the spectra were shifted to zero redshift for the individual components of the system to determine a mean and median of both the primary and secondary spectrum. A grid of model spectra was computed with the non-LTE atmosphere modelling package NGRT (Werner et al. 2003) and fitted to the median primary spectrum. The best fit so far is obtained for a model with $T_{\text{eff}} = 72\,500$ K, $\log g = 7.20$ and abundances (by number) of C/He=0.07, N/He=0.01, O/He=0.01, implying a mass fraction (He/C/N/O)=(0.78/0.16/0.03/0.03), typical for PG 1159 stars − see e.g. Werner & Herwig (2006). Typical errors for this kind of analysis are $\pm 10\,000$ K for T_{eff} and ± 0.3 for $\log g$ (Ziegler et al. 2009). The temperature obtained is significantly lower than the one found by Nagel et al. (2006), who obtained the preliminary values $T_{\text{eff}} = 90\,000$ K and $\log g = 7.60$, and implies that SDSS J212531.92−010745.9 is one of the coolest PG 1159 stars.

To get the stellar parameters of the irradiated secondary, a second model grid is being calculated using PHOENIX (Hauschildt et al. 1997). Up to now, no quantitative result has been obtained, but the Balmer series of hydrogen in emission is reproduced by the first models calculated.

Three seasons of photometry are available for SDSS J212531.92−010745.9. The system shows flux variations with a peak-to-peak amplitude of about 0.7 mag and a period of about 6.96 h. The ephemeris could be determined to a high accuracy. The light curve profile obtained shows no eclipse and is currently being fitted with the binary modelling programs nightfall and PHOEBE (Prša & Zwitter 2005). This light curve modelling possibly constrains the inclination of the system which is needed to deduce the mass of the components from the RVs. Together with the constraints of an optimized spectral analysis (especially with PHOENIX) this will give a possible mass range for the PG 1159 component of SDSS J212531.92-010745.9 (work in progress: Beeck 2009, Schuh et al. 2009, and references therein).

Acknowledgments. The spectroscopy is based on service observations collected by J. Aceituno and U. Thiele at the Centro Astronómico Hispano Alemán, operated jointly by the Max-Planck Institut für Astronomie and the Instituto de Astrofísica de Andalucía. Special thanks to B. Gänsicke and M. Schreiber who first directed our attention to this unique object, and to all observers for the photometric observations. We also thank the Astronomische Gesellschaft as well as the conference sponsors and in particular HELAS for financially supporting the poster presentation at JENAM 2008 through travel grants to B.B. and S.S.

References

Beeck, B. 2009, Diploma thesis, University of Göttingen, in prep.

Hauschildt, P., Baron, E., & Allard, F. 1997, ApJ, 490, 803

Nagel, T., Schuh, S., Kusterer, D.-J., et al. 2006, A&A, 448, L25

Prša, A., & Zwitter, T. 2005, ApJ, 628, 426

Schuh, S., Beeck, B., & Nagel, T. 2009, in "White Dwarfs", J. Phys.: Conf. Ser., in press, arXiv:0812.4860

Werner, K., Deetjen, J. L., Dreizler, S., et al. 2003, in "White Dwarfs", NATO ASIB Proc., 105, 117

Werner, K., & Herwig, F. 2006, PASP, 118, 183

Ziegler M., Rauch, T., Werner, K., Koesterke, L., Kruk, J. W., 2009, CoAst, 159, 107

Comm. in Asteroseismology,
Vol. 159, 2009, JENAM 2008 Symposium № 4: Asteroseismology and Stellar Evolution
S. Schuh & G. Handler

TT Arietis – observations of a Cataclysmic Variable Star with the MOST Space Telescope

J. Weingrill,[1] G. Kleinschuster,[2] R. Kuschnig,[3] J. M. Matthews,[4] A. Moffat,[5] S. Rucinski,[6] D. Sasselov,[7] and W. W. Weiss[3]

[1] Space Research Institute, Austrian Academy of Sciences, Schmiedlstrasse 6, 8042 Graz, Austria
[2] Astro Club Auersbach, Wetzelsdorf 33, 8330 Feldbach, Austria
[3] Institut für Astronomie, Universität Wien, Türkenschanzstrasse 17, 1180 Wien, Austria
[4] Department of Physics & Astronomy, University of British Columbia, 6224 Agricultural Road, Vancouver, BC V6T 1Z1, Canada
[5] Département de physique, Université de Montréal C.P. 6128, Succ. Centre-Ville, Montréal, QC H3C 3J7, Canada
[6] David Dunlap Observatory, Department of Astronomy, University of Toronto P.O. Box 360, Richmond Hill, ON L4C 4Y6, Canada
[7] Harvard-Smithsonian Center for Astrophysics, 60 Garden Street, Cambridge, Massachusetts, MA 02138, USA

Abstract

We measured the photometric flux of the cataclysmic variable TT Arietis (BD+14 341) using the MOST space telescope. Periodic oscillations of the flux reveal the photometric period as well as other features of this binary system. We applied a Discrete Fourier Transform (DFT) on a reduced data set to retrieve the frequencies of TT Arietis. The analysis of the system revealed a primary photometric period of 3.19 hours. Though the MOST data has a high cadence of 52.8 seconds, a fine structure of the accretion disk is not obvious.

Individual Objects: TT Ari

Introduction

The MOST (Microvariability and Oscillations of STars) satellite observed TT Ari between MJD 54395.6 and 54406.4. The optical setup of MOST consists of a Maksutov type optical telescope and two identical CCDs. A detailed description of the MOST mission can be found in Walker et al. (2003) and Matthews (2004). The classification of the star is hindered by the low inclination of 20 degrees and therefore mentioned differently in literature. It is most likely to be a VY Scl-type star (Wu et al. 2002) or belongs to the class of SW Sex stars as mentioned by Kim et al. (2009). The primary photometric period varies between 3.1824(48) hours and 3.19056(72) hours as listed in Tremko et al. (1996). According to the observations of quasi-periodic oscillations (QPO) between 2005 and 2006, TT Ari is believed to return from its 'positive superhump' state (Kim et al. 2009).

Methods

The target was measured with direct imaging photometry because of the low magnitude of the TT Ari system. The initial data reduction has been accomplished by the MOST science

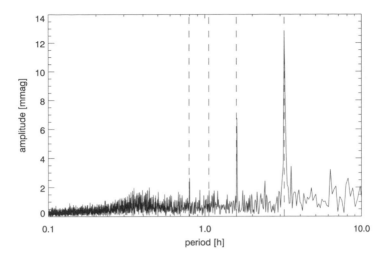

Figure 1: Periodogram derived from the DFT analysis. The main period and its harmonics are indicated by dashed lines.

team (Rowe et al. 2006). The first analysis of the data was carried out using discrete Fourier transform (Swan 1982). The Fourier autocorrelation function (Scargle 1989) was used as an alternate method, since it shows 36 periods which are not obvious in the original data. In order to remove the remaining signature of the satellite orbit, the first 100 data points equivalent to a time lag of 1.452 hours were ignored. The distances between the first three local maxima were averaged to obtain the period from the autocorrelation function. In order to look for QPOs and variations in the periodicity of the signal indicated by the autocorrelation function, a sliding Fourier window analysis (Jacobsen & Lyons 2003) was carried out. The Fourier window sizes were scaled from 1.88 hours to 15 hours and a boxcar function was used as a windowing function. The window was shifted over the data points without overlapping.

Results

The analysis using DFT reveals a period of 3.177(44) hours with an amplitude of 12.4 mmag. The resulting periodogram can be seen in Fig. 1. The main period and the amplitudes of its harmonics correspond to a nearly sinusoidal shape of the folded light curve. Other methods like the autocorrelation of the data show a slightly different photometric period of 3.194(18) hours.

The QPOs which can be identified as flickering of the accretion disk is visible in the regime of 19 to 26 minutes. This corresponds to earlier results (see e.g. Semeniuk et al., 1987) in a 'negative superhump' state. Due to the short observation run of the MOST satellite the current 'superhump state' could not be verified.

Acknowledgments.   Data was provided by 'Universe in a Suitcase − MOST for all'.

References

Walker, G., Matthews, J. M., Kuschnig, R., et al. 2003, PASP, 115, 1023

Matthews, J. M., 2004, Bullet. of the AAS, 35, 1563

Wu, X., Li, Z., Ding, Y., et al. 2002, ApJ, 569, 418

Kim, Y., Andronov, I. L., Cha, S. M., et al. 2009, A&A, in press, arXiv:0810.1489

Tremko, J., Andronov, I. L., Chinarova, L. L., et al. 1996, A&A, 312, 121

Rowe, J. F., Matthews, J. M., Seager, S., et al. 2006, ApJ, 646, 1241

Swan, R. R. 1982, AJ, 87, 1608

Scargle, J. D. 1989, ApJ, 343, 874

Jacobsen, E., & Lyons, R. 2003, SP-M, 74

Semeniuk I., Schwarzenberg-Czerny A., Duerbeck H., et al. 1987, A&A, 37, 197

The conference venue: main building of the University of Vienna; and reception and registration at the University Observatory.

Comm. in Asteroseismology,
Vol. 159, 2009, JENAM 2008 Symposium № 4: Asteroseismology and Stellar Evolution
S. Schuh & G. Handler

A progress report on the study of the optical variability of the old nova HR Del

I. Voloshina,[1] M. Friedjung,[2] M. Dennefeld,[2] and V. Sementsov[1]

[1] Sternberg Astronomical Institute, Moscow State University, Moscow, Russia
[2] Institut d'Astrophysique de Paris, CNRS, Paris, France

Abstract

The cataclysmic variable and old nova HR Del has been studied using ground-based photometry and spectroscopy over a period of more than 6 years. Some long term variations have been found, but no clear correlation established up to now between photometric, and spectroscopic changes.

Individual Objects: Old nova HR Del

With the aim of finding more indications of continuing activity of the old nova HR Del, new spectral and photometric observations have been obtained between 2002 and 2008, with however an irregular time sampling. Results from a preliminary study were presented by Friedjung et al. (2005). Here we present mainly the results of our complete photometric observations of this object.

HR Del (or Nova Del 1967) was a bright classical nova with however unusual properties. It brightened in July 1967 to a magnitude near $5^m.5$ from pre-outburst magnitude of about 12^m. After remaining 5 months near this magnitude, it brightened to a maximum $m_v \sim 3.5$, followed by an extremely slow irregular decline. The unusual nature of this nova might be due to the presence of a wind which was optically thin in the continuum before maximum, and whose velocity decreased during that stage, unlike that of the optically thick winds normally seen after maximum of classical novae (Friedjung 1992). The mass of the white dwarf component is low $- (0.55 \div 0.75)M_\odot -$, while the orbital period is $0^d.214165$ (Kürster & Barwig 1988). HR Del was unusually bright both before and long after its outburst (Selvelli & Friedjung 2003).
Two kinds of photometric observations were made:

- To construct the overall light curve, daily estimates of HR Del's magnitude were made with the UBV photometer on the 60-cm telescope of the Sternberg Astronomical Institute in Crimea (time resolution 10 s). They covered the years 2001-2006, with more than 1000 measurements in 3 bands.

- To study the short-term variability of HR Del, observations with an Ap47 CCD detector on the same telescope were obtained in 2005-2006, covering 23 nights, and containing more than 3500 measurements, with a time resolution of 30 s in the V band.

- High-resolution spectroscopic observations of the region of the $H\alpha$ line were obtained with the 152-cm telescope of the Haute Provence Observatory.

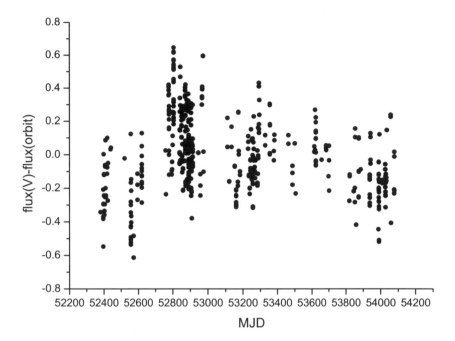

Figure 1: Long time scale flux variation of HR Del: flux in the V band corrected for orbital variations versus time in Modified Julian Day.

The overall light curve of HR Del was obtained from photometric observations in the V band and is shown in Fig. 1 after correction for orbital variations. This figure shows that, after subtraction of the orbital variations, long time scale variations exist in the light curve of HR Del. While there is evidence for long time scale variations, no short term periodic oscillations were found in the light curves determined from the CCD observations, except the orbital ones (with an amplitude up to $0^{m}5$). The profiles of Hα show both a strong narrow component and weak wings. No clear evidence was found for orbital variations of the equivalent width, but long time scale variations are seen with a decrease from 2002 to 2004, followed by an increase again. More details will be presented in a forthcoming paper.

Acknowledgments. I.V. acknowledges financial support provided by RFBR through grant 06-02-16411. We also thank N. Metlova for her kind help with observations.

References

Friedjung, M. 1992, A&A, 262, 487

Friedjung, M., Bruch, A., Voloshina, I., et al. 2005, ASP Conf. Ser., 330, 437

Selvelli, P., & Friedjung, M. 2003, A&A, 401, 297

Kürster, M., & Barwig, H. 1988, A&A, 199, 201

Comm. in Asteroseismology,
Vol. 159, 2009, JENAM 2008 Symposium № 4: Asteroseismology and Stellar Evolution
S. Schuh & G. Handler

A study of the atmospheric structure of AX Mon (HD 45910)

A. Antoniou,[1] E. Danezis,[1] E. Lyratzi,[1,2] L.Č. Popović,[3] M. S. Dimitrijević,[3] E. Theodosiou,[1]
and D. Stathopoulos[1]

[1] University of Athens, Faculty of Physics, Department of Astrophysics, Astronomy and Mechanics,
Panepistimioupoli, Zographou 157 84, Athens, Greece
[2] Eugenides Foundation, 387 Sygrou Av., 17564, Athens, Greece
[3] Astronomical Observatory of Belgrade, Volgina 7, 11160 Belgrade, Serbia

Abstract

In this paper we apply the GR model to find kinematic parameters (radial, rotational and random velocities) as well as FWHM, the absorbed energy and the Gaussian Typical Deviation (σ) for a group of FeII spectral lines from AX Mon spectra obtained with IUE. In order to find possible stratification in the FeII absorbing region of AX Mon we present these parameters as a function of the excitation potential of the lines. We found that the obtained parameters are not too sensitive to the excitation potential of the FeII lines. In addition, we calculate the above mentioned parameters for the AlII (λ 1670.81 Å), AlIII ($\lambda\lambda$ 1854.722, 1867.782 Å), MgII ($\lambda\lambda$ 2795.523, 2802.698 Å), FeII (λ 2586.876 Å), CII ($\lambda\lambda$ 1334.515, 1335.684 Å) and SiIV ($\lambda\lambda$ 1393.73, 1402.73 Å) spectral lines of AX Mon, and we present their relation with the ionization potential.

Individual Objects: AX Mon, HD 45910

Results and discussion

Using the Gauss Rotation (GR) model (Danezis et al. 1991, 2007) we accomplished the best fit of the AlII (λ 1670.81 Å), AlIII ($\lambda\lambda$ 1854.722, 1867.782 Å), MgII ($\lambda\lambda$ 2795.523, 2802.698 Å), FeII (λ 2586.876 Å), CII ($\lambda\lambda$ 1334.515, 1335.684 Å) and SiIV ($\lambda\lambda$ 1393.73, 1402.73 Å) spectral lines of HD 45910 (AX Mon). The complex structure of these spectral lines can be explained with Discrete Absorption components (DACs) and Satellite Absorption components (SACs, Danezis et al. 2007).

Variation of parameters as a function of the excitation potential

The radial and rotational velocities of the studied group of FeII lines show small changes as a function of the excitation potential. The radial velocities present three levels. The first level has values of about -260 km/s, the second one has values of about -125 km/s and the third one has values of about -18 km/s. These values are in agreement with the respective values found by Danezis et al. (1991). The values of the rotational velocities for all SACs are between 20 and 60 km/s. In the case of the random velocities of the ions of the studied group of FeII lines, we detected three levels of random velocities. The first level has values of about 115 km/s, the second one of about 70 km/s and the third one is about 35 km/s. The variation of the typical Gaussian deviation has the same form as the variation of the random velocities. There are also three levels of values. The first level has values of about 0.8, the

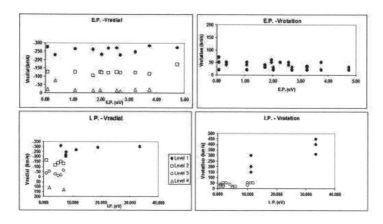

Figure 1: Radial and rotational velocities of the studied group of FeII lines as a function of the excitation potential and radial and rotational velocities of the studied group of AlII (λ 1670.81 Å), AlIII (λλ 1854.722, 1867.782 Å), MgII (λλ 2795.523, 2802.698 Å), FeII (λ 2586.876 Å), CII (λλ 1334.515, 1335.684 Å) and SiIV (λλ 1393.73, 1402.73 Å) spectral lines as a function of the ionization potential.

second one of about 0.4 and the third one of about 0.2. The Full Width at Half Maximum (FWHM, Å) of the studied group of FeII lines presents also three levels of values. The first level has values of about 2 Å, the second one of about 1.3 Å and the third one of about 0.6 Å. Finally, in the case of the absorbed energy (Ea, eV) of the studied group of FeII lines we also found three levels of values. The first level is about 1 eV, the second one about 0.4 eV and the third one about 0.14 eV.

Variation of kinematic parameters as a function of the ionization potential

Here we present the variation of the radial and rotational velocities in the AlII (λλ 1670.81 Å), AlIII (λλ 1854.722, 1867.782 Å), MgII (λλ 2795.523, 2802.698 Å), FeII (λ 2586.876 Å), CII (λλ 1334.515, 1335.684 Å) and SiIV (λλ 1393.73, 1402.73 Å) spectral lines as a function of the ionization potential. We detected four levels of radial velocities. The first level has values of about −260 km/s and corresponds to an ionization potential larger than 20 eV. The second level has values of about −140 km/s, the third one of about −35 km/s and the fourth one of about 119 km/s. All these values correspond to ionization potential with values between 0 and 10 eV. The values of the rotational velocities are 150 − 450 km/s and correspond to ionization potentials larger than 10 eV. The low values of the rotational velocities (10 − 50 km/s) correspond to ionization potentials with values between 0 and 10 eV.

Acknowledgments. This research project is progressing at the University of Athens, Department of Astrophysics, Astronomy and Mechanics, under the financial support of the Special Account for Research Grants, which we thank very much. This work also was supported by the Ministry of Science and Technological Development of Serbia, through the projects "Influence of collisional processes on astrophysical plasma line shapes" and "Astrophysical spectroscopy of extragalactic objects".

References

Danezis, E., Theodosiou, E., & Laskarides, P. 1991, Ap&SS, 179, 111

Danezis, E., Nikolaidis, D., Lyratzi, E., et al. 2007, PASJ, 59, 827

Comm. in Asteroseismology,
Vol. 159, 2009, JENAM 2008 Symposium № 4: Asteroseismology and Stellar Evolution
S. Schuh & G. Handler

η Carinae - The outer ejecta

K. Weis

Astronomisches Institut der Ruhr-Universität Bochum, Universitätsstr. 150, 44780 Bochum, Germany

Abstract

η Carinae is a unique object among the most massive evolved stars in the LBV phase. The central object(s) is(are) surrounded by a complex circumstellar nebula ejected during more than one eruption in the 19th century. Beyond the well-defined edges of its famous bipolar nebula, the Homunculus, are additional nebulous features referred to as the outer ejecta. The outer ejecta contains a large variety of structures of very different sizes and morphologies distributed in a region 0.67 pc in diameter. Individual features in the outer ejecta are moving extremely fast, up to 3200 km/s, in general the expansion velocities are between 400-900 km/s. A consequence of these high velocities is that structures in the outer ejecta interact with the surrounding medium and with each other. The strong shocks that arise from these interactions give rise to soft X-ray emission. The global expansion pattern of the outer ejecta reveals an overall bipolar distribution, giving a symmetric structure to its morphologically more irregular appearance. The long, highly collimated filaments, called strings, are particularly unusual. The material in the strings follows a Hubble-flow and appears to originate at the central object.

Individual Objects: η Car, AG Car, HR Car, P Cyg

The outer ejecta - morphology, kinematics and X-ray emission

Historic background

Van den Bos (1938) noticed for the first time extended nebular emission around η Car, Gaviola (1950) published the first images and due to its human-like appearance called the nebula the *Homunculus*. He also identified some bright regions around the Homunculus. Thackeray (1949) sketched the nebula and speculated also about a fainter outer nebula or shell. In 1976 Walborn identified structures outside the Homunculus, he called them N, E, S and W condensations, S ridge and W arc.

The morphology of the nebula

Today HST images show that η Car is surrounded by a bipolar inner nebular, the Homunculus, which consists of two lobes and an equatorial disk, its longest axis is 19″ (0.2 pc). Further out a larger clumpy irregularly shaped outer nebulae with a diameter of at least 60″ (0.67 pc) is visible and called the *outer ejecta*. Individual structures in the outer ejecta can be described as filaments, jets, arcs, bullets or knots and strings. The electron density in the outer ejecta is about $10^4 \mathrm{cm}^{-3}$, assuming a typical electron temperature of 1400 K and a volume filling factor for the clumps, a total mass of 2-4 $M_\odot$ can be derived for the outer ejecta.

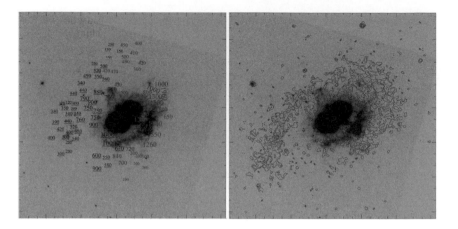

Figure 1: In grey scale an HST image of the nebula around η Car is shown. Left: Kinematic analysis shows the bipolar expansion pattern (blueshifted values underlined). Right: In contours the X-ray emission.

Kinematic analysis

Early measurements (e.g. Walborn 1976) show that expansion velocities in the outer ejecta reach up to more than 1000 km/s. Weis et al.'s (2001) systematic study revealed unexpected results for the kinematic behaviour of the individual structures with expansion velocities ranging from roughly −1000 km/s to 3200 km/s, with the majority between 400-900 km/s (see Fig. 1). While the observed features in the outer ejecta show a more or less irregular and random distribution, the red and blueshifted velocities (Fig. 1, left) show a systematic bipolar expansion pattern. Structures in the (south-)east are blueshifted, those in the (north-)west redshifted. The outer ejecta follows the same symmetry and has the same axis as the Homunculus (southern lobe tilted towards the observer, northern away).

The strings

Most amazing in the outer ejecta are long, highly collimated linear features identified by visual inspection of HST images and now labeled string 1, 2, 3, 4, 5 (Weis et al. 1999). The length of the strings varies from the shortest 0.044 pc (string 5) to 0.177 pc for the longest (string 1). The width is 0.002 to 0.003 pc, yielding a length-to-width ratio of 31 to 70. Kinematic measurements show that all strings show an increasing radial velocity towards their outer end, a perfect Hubble-flow.

X-ray emission

An overlay of an HST with a CHANDRA image (soft X-ray emission 0.6-1.2 keV) shows a very good spatial agreement between the optical and diffuse X-ray emission (Fig. 1, right). The X-ray emission exhibits a hook shape structure, composed of smaller substructures. Across the hook, a *bridge* projects onto the Homunculus from the south to the north-east. The X-ray emission of the outer ejecta is created by shocks due to collisions of the individual structures with the ISM and with each other. Determined from postshock temperatures, the postshock velocities necessary to reproduce the X-ray emission are roughly 670-760 km/s, exactly what is observed. It can be concluded that the X-ray image of the outer ejecta displays the shock fronts of the interacting material.

Other LBV nebula, some facsimiles of the outer ejecta

η Car is not the only LBV surrounded by a nebula. A statistic of known, and resolved, LBV nebulae shows that 50% of the nebulae are bipolar to some degree, about 40% are roughly spherical, and less than 10% appear completely irregular. Size and kinematics show that the nebula around HR Car appears as an older twin of the Homunculus (Weis et al. 1997). HR Car's nebula but also the nebulae around AG Car and P Cygni show larger components which could be regarded as facsimiles of the outer ejecta.

A complete review article on the outer ejecta (Weis 2009) will appear in "η Carinae and the supernovae imposters", eds. K. Davidson and R. M. Humphreys, Springer-Verlag.

Acknowledgments. I thank D.J. Bomans and W.J. Duschl for countless discussions and their input.

References

van den Bos, W. H. 1938, Union Obs. Circ., 100, 522
Gaviola, E. 1950, ApJ, 111, 408
Thackeray, A. D. 1949, The Observatory, 69, 31
Walborn, N. R. 1976, ApJ, 204, L17
Weis, K., Duschl, W. J., Bomans, D. J., et al. 1997, A&A, 320, 568
Weis, K., Duschl, W. J., & Chu, Y.-H. 1999, A&A, 349, 467
Weis, K., Duschl, W. J., & Bomans, D. J. 2001, A&A, 367, 566

Lunch break in the Arcades of the main building of the university.

Comm. in Asteroseismology,
Vol. 159, 2009, JENAM 2008 Symposium № 4: Asteroseismology and Stellar Evolution
S. Schuh & G. Handler

Time series analysis with the VSAA method

S. Tsantilas,[1] K. Kolenberg,[2] and H. Rovithis-Livaniou [1]

[1] Dept. of Astrophysics, Astronomy & Mechanics, Faculty of Physics, Athens University,
Panepistimiopolis, Zografos 157 84, Athens, Greece
[2] Institut für Astronomie, Universität Wien, Türkenschanzstrasse 17, 1180 Vienna, Austria

Abstract

Time series analysis is a common task in many scientific fields, and so it is in astronomy, too. Fourier Transform and Wavelet Analysis are usually applied to handle the majority of the cases. Even so, problems arise when the time series signal presents modulation in the frequency under inspection. The Variable Sine Algorithmic Analysis (VSAA) is a new method focused exactly on this type of signals. It is based on a single sine function with variable coefficients and it is powered by the simplex algorithm. In cases of phenomena triggered by a single mechanism — that Fourier Transform and Wavelet Analysis fail to describe practically and efficiently — VSAA provides a straightforward solution. The method has already been applied to orbital period changes and magnetic field variations of binary stars, as well as to the Blazhko effect of the pulsating RR Lyrae stars and to sunspot activity.

Individual Objects: CG Cyg, RR Lyr

Introduction

Research on astronomical data often involves time series analysis. This also holds for time series of pulsating stars showing modulation, of orbital period variations in binary systems, of sunspot or starspot activity, etc. The analysis of these time series is usually carried out using the well-known Fourier transform. The problem that arises in the majority of these cases is that the modulation of the phenomena is not strictly periodic but more of a quasi-periodic nature. Because the Fourier transform is focused on the frequency domain only, the micro-variations in amplitude and frequency through time are untraceable with this method.

The VSAA method

The VSAA — Variable Sine Algorithmic Analysis — (Tsantilas & Rovithis 2007) is an method incorporated in a computer code and powered by the Simplex algorithm (Nelder & Mead 1965). It is based on a single sine function with coefficients that are not constant but are functions of time. Instead of having a large number of frequencies needed to explain the signal (as in Fourier and Wavelet analysis), the VSAA traces the micro-variations of a single frequency more or less continuously. Therefore, the output of the analysis is simple and clear, while the results can be assigned in a straightforward manner to a single mechanism that modulates its characteristics (amplitude and frequency) through time. The core function is:

$$f(t) = a \cdot \sin(b \cdot t + c), \tag{1}$$

where $a = a(t)$, $b = b(t)$ and $c = c(t)$, i.e. they are functions of time.

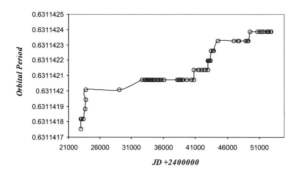

Figure 1: Orbital period variation of CG Cyg.

The user has to define some initial values for the amplitude, the starting frequency and the (possible) phase shift of the signal, the sliding window width, the number of iterations for every step of the simplex, and the accuracy threshold of the simplex.

The three major parameters: frequency, amplitude and phase shift, can be defined independently as adjustable or fixed. Because the frequency and phase shift are connected, it is suggested to fix one of them. After the initiation of the starting values the program performs a windowed partial fit using the simplex algorithm. This procedure continues automatically until the end of the time series data set under inspection. The output of the analysis is a set of $N - \ell$ vectors v_j, where N is the number of the input data points and ℓ is an internal parameter which secures that there are enough points in order to get a decent fit. The vectors have the form: $v_j = (t_j, a_j, b_j, c_j, \sigma_j, F_j), j = 1...N - \ell$, where t_j denotes the time, $a_j(t)$ is the amplitude of the signal, $b_j(t)/2\pi = f(t)$ denotes the variable frequency, $c_j(t)$ stands for the phase shift of function (1), $\sigma_j = \sqrt{\frac{\Sigma(s_j - F_j)^2}{w}}$ is the mean error, where s_j is the original signal data and F_j is the VSAA fit to the signal.

Applications

The VSAA has already been applied to synthetic data (Tsantilas & Rovithis 2008b), to the analysis of the Blazhko effect (Kolenberg & Tsantilas 2008) and to sunspot data (Reindel et al. 2008, Tsantilas & Rovithis 2008a). By fixing the amplitude to an arbitrary value, the VSAA could also trace the micro-variations of the actual period of a system in the time-frequency domain. Here we present such a first application: the extraction of the orbital period of the eclipsing binary CG Cyg directly from its times of minima (Fig. 1). It is important to notice here that the period can be acquired without the involvement of the O–C (observed minus calculated time of minima) diagram.

References

Kolenberg, K., & Tsantilas, S. 2008, CoAst, 157, 52

Nelder, J. A., & Mead, R. 1965, Comput. J., 7, 308

Reindel, A., Bradley, P., Tsantilas, S., & Guzik, J. 2008, GONG 2008/SOHO XXI conference, Boulder, Colorado

Tsantilas, S., & Rovithis-Livaniou, H. 2007, arXiv:0709.3224

Tsantilas, S., & Rovithis-Livaniou, H. 2008a, Rom. Astron. Journal, in press

Tsantilas, S., & Rovithis-Livaniou, H. 2008b, CoAst, 157, 87

Comm. in Asteroseismology,
Vol. 159, 2009, JENAM 2008 Symposium № 4: Asteroseismology and Stellar Evolution
S. Schuh & G. Handler

Mapping pulsations on rapidly rotating components of eclipsing binaries

B. I. Bíró [1] and O. Latković [2]

[1] Baja Astronomical Observatory, H-6500, Szegedi út, P.O. Box 766, Baja, Hungary
[2] Astronomical Observatory, Volgina 7, 11000 Belgrade, Serbia

Abstract

We report on the progress in the development of an eclipse mapping method for reconstruction of non-radial pulsation patterns on components of eclipsing binaries. In this paper we present our attempts to use the method on distorted modes caused by rapid rotation, and we estimate that it can properly detect this phenomenon.

Introduction

In previous publications (Bíró & Nuspl 2005, Latković & Bíró 2008) we have presented the development of the eclipse-mapping method for reconstruction of surface patterns of non-radial oscillations on pulsating stars in eclipsing binaries. So far, the method was tested on artificial light curves computed for a simple model: spherical stars, circular orbit and monochromatic radiation. It was proven that, in principle, many simultaneous non-radial oscillation modes can be reconstructed to a degree that allows mode identification, provided that enough and precise observational data are available. The goal of our most recent work is to prepare the method for application to real data. As a first step in this direction, we have tested whether the method can detect the mode distortion effect, which is expected in rapidly rotating stars.

Binarity, pulsation and rapid rotation

The effects of rapid rotation on stellar pulsations have recently been reconsidered by Ligniéres et al. (2006). Their scheme applies especially to p modes of δ Scuti stars, the most frequent type of pulsators found in eclipsing binaries. They show that the validity of the perturbative approach to analysis of pulsations becomes questionable already at modest rotational velocities of 50 km/s. δ Scuti stars generally rotate even faster, up to 200 km/s (Rodríguez et al. 2000). In addition, tidal synchronization in eclipsing binary systems also enforces large rotation on the components of any type – therefore the need to account for rotationally distorted modes.

As shown by Ligniéres et al. (2006), the most important effect of rapid rotation on a pulsation mode is the concentration of the amplitude towards the equator. This effect increases with the frequency of the mode and drastically changes its visibility in integrated light. This is good news for eclipse mapping, since modes with degrees up to $l = 4$ become photometrically detectable, and thus can be reconstructed.

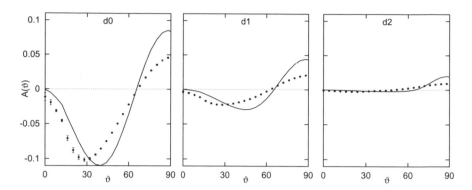

Figure 1: Reconstructed amplitude profiles of an $l = 4$, $m = 2$ mode as a function of co-latitude, for three levels of distortion (d_0 being the "no distortion" case). The model profiles are plotted with solid lines, and the reconstructed profiles with dots. Note how the response factor increases with the distortion: the three cases yield the same out-of eclipse oscillation amplitude in the light curve.

Results

We have tested the method on a number of artificial eclipsing binary systems, with different pulsation modes. For demonstration here we chose a model with stellar radii $r_1 = 0.14$ and $r_2 = 0.36$ (in units of separation), inclination of $80°$, and one single mode with $l = 4$ and $m = 2$. Since the mode distortion becomes significant even at relatively low rotational velocities, we did not alter the spherical shape of the pulsator. Having no analytical expression for the mode distortion, we used an empirical approach that roughly reproduces the equatorial amplitude concentration, since the main purpose of this study is only to check whether such a kind of deviation from spherical harmonics can be detected by the eclipse-mapping method.

Figure 1 shows the amplitude profiles on the northern hemisphere for three levels of distortion. The phase profiles (omitted for brevity), show no difference in the three cases, and are properly restored (as expected given the good longitudinal resolution of the eclipse's surface sampling), yielding the correct $m = 2$ order.

Conclusions

Our investigation shows that for all modes for which the rotational distortion produces sensible changes in the modulation of the pulsations during the eclipses, the distortion of the modes can be detected and, in principle, properly restored by the Eclipse Mapping method.

Acknowledgments.   This work was supported by Hungarian OTKA Grant No. F-69039 and by the Ministry of Science and Technological Development of Serbia through project no. 146003, "Stellar and Solar Physics".

References

Bíró, I. B., & Nuspl, J. 2005, ASP Conf. Ser., 333, 221

Latković, O., & Bíró, I. B. 2008, CoAst, 157, 330

Ligniéres, F., Rieutord, M., & Reese, D. 2006, A&A, 455, 607

Rodríguez, E., López-González, M. J., & López de Coca, P. 2000, A&AS, 144, 469

Comm. in Asteroseismology,
Vol. 159, 2009, JENAM 2008 Symposium № 4: Asteroseismology and Stellar Evolution
S. Schuh & G. Handler

Physical parameters of contact binaries through 2-D and 3-D correlation diagrams

K. D. Gazeas

Harvard-Smithsonian CfA, 60 Garden Street, Cambridge, MA 02138, USA

Abstract

Physical parameters of contact binaries are studied through 2-D and 3-D correlations among them. It is shown that the physical parameters (i.e. mass, radius and luminosity) are closely correlated with the orbital period and mass ratio in the 3-D domain. These correlations can be used as a quality check for the parameters in every given solution of a contact binary. The empirical laws, extracted out of these correlations, are a useful tool for a quick estimate of physical parameters for the numerous contact binaries found in global sky surveys.

The 2-D and 3-D correlation diagrams

Our sample is based on the list of 112 contact binaries published by Gazeas & Stepien (2008). These binaries are the only systems with accurate solutions, based on high quality photometric light curves and good radial velocity curves for both components. We show that several relations and correlations exist among the discussed parameters. Some of them may be useful in the future for approximate estimates of masses and radii of contact binaries for which the orbital period is known. In our study we consider component "1" as the presently more massive one. Our assumption is based on the double-lined spectroscopic observations, where the mass ratio is taken as $q = M_2/M_1 \leq 1$.

As shown also by Gazeas & Niarchos (2006), the plots of mass and angular momentum versus orbital period gave a direct evidence of evolution into contact, as well as the evolution of A-type W UMa binaries towards the W-type. Figure 1 (upper panel) shows a plot of mass versus orbital period, as well as the equivalent plots of radius and luminosity in logarithmic scale. It was easy to extract empirical laws out of these diagrams, as the correlation of absolute parameters with the orbital period is more than obvious, with an error of approximately 15% (Eqs. 1-6). In the lower panel, the same physical parameters are plotted against orbital period and mass ratio as: $M = M(P, q)$, $R = R(P, q)$, $L = L(P, q)$. The extracted empirical laws (Eqs. 7-12) give an error less than 5%.

Conclusions

The errors derived from the 3-D correlation equations are significantly small, especially in the prediction of mass and radius. The determination of the physical parameters from the above diagrams is independent of the orbital inclination and (possible) presence of additional objects in a system, derived from the light curve. It is so accurate that it can be used in the opposite way, checking the solutions obtained with the classic procedure. The importance of such an approximation is that we will be able to calculate directly the absolute parameters

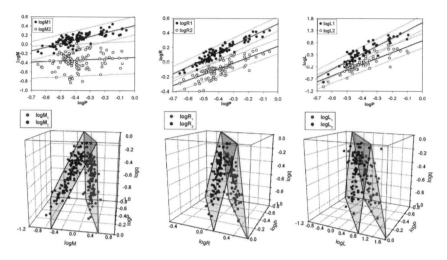

Figure 1: The 2-D (upper panel) and 3-D (lower panel) correlation diagrams for the absolute physical parameters M, R and L, as described in the text by Eqs. 1-6 and 7-12, respectively.

$$\log M_1 = 0.755(59) \cdot \log P + 0.416(24) \tag{Eq. 1}$$
$$\log M_2 = 0.352(17) \cdot \log P - 0.262(67) \tag{Eq. 2}$$
$$\log R_1 = 0.992(38) \cdot \log P + 0.533(16) \tag{Eq. 3}$$
$$\log R_2 = 0.801(63) \cdot \log P + 0.221(26) \tag{Eq. 4}$$
$$\log L_1 = 3.256(199) \cdot \log P + 1.645(82) \tag{Eq. 5}$$
$$\log L_2 = 2.444(187) \cdot \log P + 0.892(77) \tag{Eq. 6}$$

$$\log M_1 = 0.725(59) \cdot \log P - 0.076(32) \cdot \log q + 0.365(32) \tag{Eq. 7}$$
$$\log M_2 = 0.725(59) \cdot \log P + 0.924(33) \cdot \log q + 0.365(32) \tag{Eq. 8}$$
$$\log R_1 = 0.930(27) \cdot \log P - 0.141(14) \cdot \log q + 0.434(14) \tag{Eq. 9}$$
$$\log R_2 = 0.930(29) \cdot \log P + 0.287(15) \cdot \log q + 0.434(16) \tag{Eq. 10}$$
$$\log L_1 = 2.531(67) \cdot \log P - 0.512(51) \cdot \log q + 1.102(43) \tag{Eq. 11}$$
$$\log L_2 = 2.531(63) \cdot \log P + 0.352(52) \cdot \log q + 1.102(41) \tag{Eq. 12}$$

of such systems, knowing only their fundamental observational characteristics (orbital period and mass ratio). This will provide a great opportunity to study easily and much faster the properties of thousands of contact binaries that were discovered (or will be discovered soon) by the global sky surveys. On the other hand, the absolute parameters will be used as distance indicators, since contact binaries have without doubt many advantages as standard candles (Rucinski 1996). It is important to note that the 2-D and 3-D empirical laws do not give the full solution of contact binary systems. They do give a very accurate approximation of their physical parameters, as well as a very good view of the contact binary systems' structure.

References

Gazeas, K., & Niarchos, P. 2006, MNRAS, 370, 29

Gazeas, K., & Stepien, K. 2008, MNRAS, 390, 1577

Rucinski, S. M. 1996, ASP Conf. Ser., 90, 270

Comm. in Asteroseismology,
Vol. 159, 2009, JENAM 2008 Symposium № 4: Asteroseismology and Stellar Evolution
S. Schuh & G. Handler

Musical scale estimation for some multiperiodic pulsating stars

B. Ulaş

Department of Physics, Onsekiz Mart University of Çanakkale, 17100, Çanakkale, Turkey

Abstract

The agreement between frequency arrangements of some multiperiodic pulsating stars and musical scales is investigated in this study. The ratios of individual pulsation frequencies of 28 samples of various types of pulsating stars are compared to 57 musical scales by using two different methods. The residual sum of squares of stellar observational frequency ratios is chosen as the indicator of the accordance. The result shows that the arrangements of pulsation frequencies of Y Cam and HD 105458 are similar to Diminished Whole Tone Scale and Arabian(b) Scale, respectively.

Individual Objects: Y Cam, HD 105458

Introduction

Musical scales are made of several musical notes. Since every note is defined by a typical definite frequency value, it is possible to say that a musical scale is a group of frequencies, listed from the low frequency to high, in one octave. Although every scale has its mathematical representation, their creation and evolution strongly depend on social and cultural structure of the societies as well as the traditional musical instruments.

Multiperiodic pulsating stars show more than one pulsation frequency. They are generally classified according to their pulsational behaviours, shape of light curves and pulsation periods. The frequencies can be obtained by using photometric or spectroscopic methods and analyses. These analyses provide information about the interiors of the stars which plays an important role in astrophysics.

To investigate the agreement between frequency arrangements of some multiperiodic pulsating stars and musical scales, we first arranged the stellar frequencies from low to high frequency in order to make the investigation meaningful and the data comparable. Then we had two data sets - one for musical scale and one for the star - so as to apply the process. The residual sum of squares (RSS) of stellar observational frequencies from the musical ones was chosen as the indicator of the agreement. The probability (P) yielded from the *t-test* for these two data sets was also calculated (Salkind 2006). We compared 57 musical scales (Slonimsky 1986, Berle 1997) which can be classified in 9 scale groups[1] (Bebop, Blues, Exotic, Harmonic, Major, Melodic Minor, Pentatonic, Symmetric, other scales) to 28 pulsating stars of β Cep, δ Sct, γ Dor, oEA, roAp, RV Tau, sdB, solar-like, SPB, and ZZ Cet type[1].

---

[1] The complete list of musical scales and stars used in this study is available online from http://bulash.googlepages.com/list

Table 1: The results of the calculation for Y Cam (oEA type) and HD 105458 (γ Dor type). S1 and S2 refer two different calculation types described in the text. The sample song given in the last column uses estimated scale - partly or completely.

Calculation	Star	Scale	RSS	P	Sample
S1	Y Cam	Diminished Whole Tone	0.003	0.861	1
	HD 105458	Lydian	0.007	0.974	2
S2	Y Cam	Spanish	0.005	0.689	3
	HD 105458	Arabian	0.003	0.776	4

1 Antonio C. Jobim - The Girl from Ipanema
2 Joe Satriani - Flying in a Blue Dream
3 Pink Floyd - Matilda Mother
4 Nikolas Roubanis - Misirlou

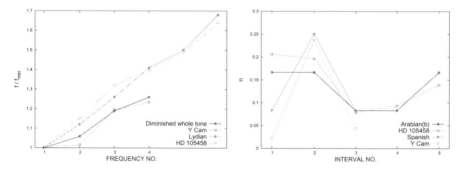

Figure 1: The best RSS value is reached for Y Cam (left) and HD 105458 (right) with the Diminished Whole Tone Scale and Arabian(b) Scale, respectively.

Calculations

Musical scales are built based on the ratios of frequencies that the human ear is sensitive to. The simple fractions, e.g., 2:1, 3:2, 4:3 are said to be consonant. We first calculated the ratio of the frequency of a given note to the lowest frequency of the scale. For instance, the Blues Scale has the frequency ratios $1 - 1.19 - 1.33 - 1.41 - 1.50 - 1.78 - 2$. Accordingly, as an additional example, the same sequence for a star, say HD 209295 (Handler et al. 2002), is $1 - 1.03 - 1.56 - 1.99 - 2.03 - 2.28$, from the low frequency to high. The best results of the calculation are shown in S1 part of Table 1 and plotted in Fig. 1.

Additionally, we compared the intervals which are ratios between consecutive frequencies. The formula is $n = \log_2(f_{i+1}/f_i)$ (Wood 2007), where n and f refer to the interval and the frequency value, respectively. The indices, i and $i+1$, denote two consecutive frequencies, for example consonant tones: $n = 0$ for unison (1:1), $n = 1$ for octave (2:1), and $n = 0.58$ for fifth (3:2). The Arabian(b) Scale has intervals for the first six notes $0.17 - 0.17 - 0.08 - 0.08 - 0.17$, while these for the star HD 105458 (Henry et al. 2001) are $0.21 - 0.19 - 0.08 - 0.09 - 0.14$. The results are listed in S2 part of Table 1 and plotted in Fig. 1. The numbers in the x-axis represent the location of the frequencies used, e.g. 1 denotes the interval of the first and the second frequencies of the group, while 2 denotes the same for second and third one.

Conclusion

The best RSS value is 0.003 for Y Cam with Diminished Whole Tone Scale estimation in the first calculation while the investigation of intervals resulted in the same value for HD 105458 with the Arabian(b) Scale estimation. Some stars can be estimated by more than one scale because of the fact that some scales contain the same notes in their first four or five orders.

References

Berle, A. 1997, Mel Bay's Encyclopedia of Scales, Modes and Melodic Patterns, Pacific, Mel Bay Publ.

Handler, G., Balona, L. A., Shobbrook, R. R., et al. 2002, MNRAS, 333, 262

Henry, G. W., Fekel, F. C., Kaye, A. B., et al. 2001, AJ, 122, 3383

Salkind, J. 2006, Statistics for People Who (think they) Hate Statistics, California, Sage Publications

Slonimsky, N. 1986, Thesaurus Of Scales And Melodic Patterns, New York, Macmillan Publishing Co

Wood, A. 2007, The Physics Of Music, Davies Press

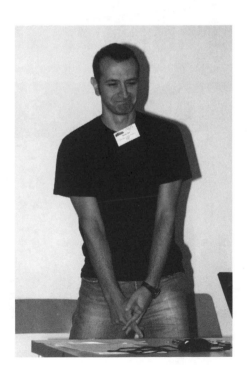

Special thanks to the photographers who kindly provided the snapshots included in this volume: Denise Lorenz and Alfred Omann.